SPECULATIONS

by

J. H. Hacsi

If it is the truth what does it matter who says it. Anon.

CHAMPAGNE PRESS
P.O. Box 5159
Whittier, CA. 90607-5159

ISBN 0-9612146-0-0

To my Mother

who is the greatest

and to Pat

ACKNOWLEDGMENTS

My thanks to the following authors, associations, periodicals and publishers for the information each supplied; also, where applicable, for graciously granting permission for me to reprint copyrighted material:

AMERICAN ACADEMY OF OPHTHALMOLOGY

AMERICAN CANCER SOCIETY

AMERICAN PSYCHOLOGICAL ASSOCIATION. Citations from PSYCHOLOGICAL ABSTRACTS are reprinted with permission of the American Psychological Association, publisher of PSYCHOLOGICAL ABSTRACTS and the PsycINFO Database (Copyright © by the American Psychological Association), and may not be reproduced without its prior permission.

ARTHRITIS FOUNDATION

THE BALTIMORE SUN

W. H. Bates, THE CURE OF IMPERFECT SIGHT BY TREATMENT WITHOUT GLASSES, Central Fixation Publishing Co., New York City: 1920.

Thomas R. Blakeslee, author of THE RIGHT BRAIN, Anchor Press/Doubleday, Garden City, New York, Copyright © 1980.

David Bohm & B. Hiley, "On the Intuitive Understanding of Nonlocality as Implied by Quantum Theory," FOUNDATIONS OF PHYSICS, Vol. 5 (1975).

CBS SONGS

Fritjof Capra. Excerpts reprinted by special arrangement with Shambhala Publications, Inc., 1920 13th Street, Boulder, Colorado 80302. From THE TAO OF PHYSICS, Copyright © 1975 by Fritjof Capra.

G. Chedd, "Genetic Gibberish in the Code of Life," SCIENCE 81, Vol. 2, No. 9, November 1981.

Paul de Kruif, MICROBE HUNTERS, Harcourt Brace Jovanovich, Inc. New York, 1926.

Phil Donahue, DONAHUE, MY OWN STORY, Simon and Schuster, New York: Copyright © 1979.

F. Eberson, MICROBE'S CHALLENGE, Ronald Press, 1963.

ENCYCLOPEDIA AMERICANA, Grolier, Inc., Danbury, Conn.

ENCYCLOPAEDIA BRITANNICA, Chicago, Ill.

Carol Eron. Excerpts reprinted with permission of Macmillan Publishing Company from VIRUS THAT ATE CANNIBALS, by Carol Eron. Copyright © 1981 by Carol Eron.

Arthur S. Freese, HELP FOR YOUR ARTHRITIS AND RHEUMATISM. Copyright © 1978 by Arthur S. Freese. Excerpts reprinted by arrangement with The New American Library, Inc., New York, New York.

H. Geist, Ph.D., THE PSYCHOLOGICAL ASPECTS OF RHEUMATOID ARTHRITIS, 1966. Excerpts reprinted courtesy of Charles C. Thomas, Publisher, Springfield, Illinois.

Philip Goldstein, GENETICS IS EASY, Lantern Press. Excerpts reprinted courtesy of the author.

Lucien Israel, MD, CONQUERING CANCER, translated by Joan Pinkham, Copyright © 1978 by Lucien Israel. Excerpts reprinted courtesy of Random House, Inc.

Edna St. Vincent Millay, for a marvelous line from LETTERS OF EDNA ST. VINCENT MILLAY, Harper and Row, Copyright © 1952 by Norma Millay Ellis. Quoted courtesy of Norma Millay Ellis.

Robert E. Netterberg, DDS, and Robert T. Taylor, THE CANCER CONSPIRACY, Copyright © 1981 by Robert E. Netterberg, DDS, and Robert T. Taylor. Excerpts reprinted courtesy of Pinnacle Books, Inc., New York, New York.

THE POCKET DICTIONARY OF AMERICAN SLANG, complied and edited by Harold Wentworth and Stuart Berg Flexner. Copyright © 1960, 1967 by the Thomas Y. Crowell Company.

RANDOM HOUSE COLLEGE DICTIONARY REVISED EDITION, Copyright © 1975, 1983 by Random House, Inc. Definitions throughout the text are from this source, reprinted courtesy of Random House, Inc.

Robert Reid, MICROBES AND MEN, Saturday Review Press: E. P. Dutton and Co., Inc. Copyright © 1974 by Robert Reid.

J. I. Rodale and Staff. Excerpts reprinted from ENCYCLOPEDIA OF COMMON DISEASES, Copyright © 1962 by J. I. Rodale. Permission granted by Rodale Press, Inc., Emmaus, PA. 18049.

W. A. Rossi, THE SEX LIFE OF THE FOOT AND SHOE, Saturday Review Press: E. P. Dutton and Co., Inc. Copyright © 1976 by W. A. Rossi.

Maggie Scarf, "Images That Heal," PSYCHOLOGY TODAY, Sept. 1980.

C. Stern, PRINCIPLES OF HUMAN GENETICS, W. H. Freeman and Co., San Francisco: Copyright © 1973 by C. Stern.

Greer Williams, VIRUS HUNTERS, Alfred A. Knopf, Inc. Copyright © 1959 by Greer Williams.

W. B. Yeats. For two provocative lines from "Crazy Jane Talks with the Bishop." Reprinted with permission of the Macmillan Publishing Company from COLLECTED POEMS of W. B. Yeats, Copyright 1933 by Macmillan Publishing Co., Inc., renewed 1961 by Bertha Georgie Yeats.

UNIVERSITY MICROFILMS INTERNATIONAL, publishers of DISSERTATION ABSTRACTS INTERNATIONAL.

I also wish to thank Dr. O. Carl Simonton, Fort Worth, Texas, who, in dealing with cancer, had the courage to look beyond the primitive demon theory of disease; and Andrew J. Galambos, Director, The Free Enterprise Institute, Los Angeles, CA., who stresses the need to define one's terms, and who also stresses the need to take risks.

TABLE OF CONTENTS

1

THE HUMAN BODY

Did Some One or Some Force Have Something in Mind?

What a piece of work is a man!
Shakespeare

You who are reading these words almost certainly inhabit a body. Chances are that the body is human. Have you ever given thought to the physical form in which you find yourself living?

Most of us have become so accustomed to the bodies in which we are tenants, are so used to being surrounded by other similar bodies, we are not apt to give much thought to the architecture of our forms. As long as our bodies possess the standard equipment and this equipment functions reasonably well, we ordinarily enjoy the physical pleasures the body affords us—thinking, talking, singing, seeing, hearing, touching, walking, running, jumping, dancing, eating, drinking, making love—without much thought. The closest we may come to thought is a mild annoyance over real or imagined imperfections: our body is too tall or too short, carries too much weight or not enough, is under-endowed or over-endowed in certain areas we may consider vital; but apart from this we look without seeing, accept without questioning, take our physical selves for granted.

But consider this:

We are here. In some fashion or other we arrived here and now live in the particular bodies we have. Possibly this all happened by accident. On the other hand, possibly it wasn't by accident. Possibly some Being (God or Allah or Whoever) and/or some Force (evolutionary, revolutionary or whatever) created the forms we know today.

Just for fun, let's set aside for the moment the possibility that we might simply be here by accident, and consider the alternate possibility: that we are not simply the end result of some meaningless accident, but rather that somewhere in our misty background there existed, or still exists, a Being and/or Force responsible for our creation.

If this Being and/or Force created the forms we know today, then surely one of the most pertinent questions we can ever ask is: in creating the human body, did that Being and/or Force have something specific in mind?

Were our bodies carefully fashioned according to some as yet little understood blueprint to achieve some specific purpose?

If our astronauts landing on the moon in July 1969 had found there a strange machine, the purpose of which they could not immediately comprehend, they would surely have studied it minutely, part by part as well as the interrelations between the various parts, in an attempt to determine for what purpose it had been constructed.

If we wonder why we are here, wonder about the purpose—or, as some see it, lack of purpose—of human life on earth, why not subject our bodies to the same scrutiny with which we would study some alien machine to see if by any chance some clues as to our purpose have been built into our bodies?

Science has taught us a great deal about the *how* of our bodily operation, but little about the *why* of our construction.

Why is the human body formed as it is?

Why is it supplied with two lungs, two kidneys, two legs, two arms, two eyes, two ears, two testicles or ovaries, yet only one heart, one brain, one nose, one stomach and one liver?

Why was the heart placed toward the left rather than to the right or at a tidy, non prejudicial dead center?

Why does the female body have a much more pronounced waist than the male body?

Why is the male body customarily far more hairy than the female body?

Why was the body so formed that for most of us sex is highly pleasurable

But Love has pitched his mansion in
The place of excrement.

Yeats

Is there some purpose in this placement or were we simply designed by a Someone too feeble-minded to do better by us aesthetically? or by a Force indifferent to all aesthetic considerations?

Why, in the male body, is one organ, the penis, called upon to function for the discharge of both urine and semen, thereby forcing a choice between the two functions?

Why are some female bodies so formed that if a tampon has been inserted into the vagina, a bowel movement will eject it? The vagina and rectum are of course proximate to each other, but so are those two other body orifices, the mouth and the nose, yet you can spit without causing mucus to trickle from the nose or blow the nose without automatically expelling spittle.

Why are we so built that, to remain healthy, we must periodically eject solid matter in phallic form? And why, in this day and age, when psychology talks freely of phallic objects, is this particular phallic look-alike, the well-formed stool, rarely or never called to our attention? Very young children are aware of the resemblance and find meaning in it, but who remembers it clearly once childhood has passed?

To answer any or all of the above questions by responding, "Because. Because that's the way man has developed. Because that's the way life is and we share these, or some of these, characteristics with all other animals, or all other vertebrates," is to beg the question. It's as though we were to answer the query, "Why does the Chevrolet automobile have four wheels and two headlights?" with the retort, "Because the Ford car does." At some point in time, back before the first automobile was ever built, there was a planning stage in which it was decided to construct the automotive machine with not three or five wheels but four and not one or three headlights but two, and the descendents of that original car now speed across our land with the predetermined number of wheels and headlights.

For the sake of argument, let's assume for the moment that man too, like the automobile, is the product of a creative intelligence, a manufactured product almost the rival of the motorcar in importance, and that sometime back in the pre-production, drawing board stage for the animal life due to evolve upon the planet earth, there was similar planning during which it was decided that life would develop along certain lines, for certain specific reasons, to ensure that Homo sapiens would, in time, emerge in triumphant, ideal form.

What is this triumphant, ideal form as we see it today?

A parts list for the human form would go something like this:

PARTS LIST – HUMAN

SKELETAL FRAMEWORK, PADDING, ET CETERA:

Bones (200-212), long/flat, various sizes
Marrow
Tissue: connective (cartilage, tendon, ligament)
muscle, fatty, fibrous, lymphoid, membrane,
epithelial
Neurons

FLUIDS:

Lubricating
Lymph
Blood

ASSORTED GLANDS:

One Assortment "His"
One Assortment "Hers"

INTERIOR AND SENSORY ORGANS AND APPENDAGES:

One Each	Two Each
Heart with fittings*	Eye
Brain with fittings	Ear
Nose with fittings**	Arm with hand/fingers
Stomach with fittings***	Leg with foot/toes
Liver with attachment****	Lung with fittings
Penis or Uterus with fittings	Kidney with fittings*****

*Include arterial/venous network complete with capillaries

**Include vocal chords in trachea

***Include tongue with taste buds in oral cavity; also include spleen; also intestines, small, 21 feet approx.; intestines, large, 6 feet approx.; outlet; verniform appendix.

****Gall bladder

*****Include ureters, bladder, uretha.

ACCESSORIES:

Teeth, Nails, Hair

EXTERIOR WRAPPING:

Skin (epidermis, corium)

Proceeding still under the assumption (unproved to many of us, perhaps not scientifically provable ever) that we are not here as a result of a meaningless accident or a series of meaningless accidents, but are instead the product of a creative intelligence—either a Being Who created the human body substantially as it exists today and/or a Force Which determined in advance its evolutionary development into what it is today—let's inspect this parts list to see if any interesting clues present themselves.

First, as when we go to buy an automobile, we can, if we wish, dispense with the accessories. Accessories may spruce up a motor machine, make it more eye catching or easier to operate, but even without them the car will function. Man's accessories too, regardless of their importance in past eras and the convenience of still possessing them, can be dispensed with without loss of vital function. We can lose our hair, our nails, even our teeth without loss of life or serious threat to our well being.

If we look at ourselves we will see how generous God or Nature was with these—in our times—decorative non essentials. The mature human being is supplied with a standard set of thirty-two teeth. Lions and tigers, who would have a difficult time surviving in the wild without their teeth, have to struggle along with thirty, two fewer than we have. There has never been a time since man became man that he couldn't live without teeth—yet we are very generously supplied with them.

When it comes to nails, we are also very generously supplied. Counting finger and toe, we have twenty nails per body when we can survive with none. This is a surplus of two over the claws that the ferocious big cats have. As for bodily hair, which we can very easily survive without, we have such a surplus that the market is filled with products designed to help both sexes rid themselves of excess. How munificently the Being Who or the Force Which designed us treated its prize creation, Man.

But wait a minute. Have we really been so generously supplied? When it comes to embellishments, yes. We have been very amply supplied with what we can live without. As far as the human body goes, multiplicity seems almost to equal dispensable.

What about those anatomical parts more essential to life?

Let's break down the **PARTS LIST – HUMAN** into those parts which are currently essential to survival and those which are not essential to survival, noting how many of each comes as standard equipment.

Note: Medical science has now progressed to the point where machines can take over and perform the function of almost every organ in the body and in this way sustain "life." Also, organ transplants occur, with damaged organs being replaced. In the discussion which follows, these miracles of modern science are ignored. References to the continuation or non continuation of life refer to self sustaining life, life in the absence of machine takeover or surgical replacement of parts. However, the same principles apply even when these medical advances are taken into account, i.e., man cannot live without breathing, whether the breathing function is accomplished by his own lungs or a machine. Similarly with all other vital functions.

PARTS LIST – HUMAN (Partial)

Life Depends Upon	Life Does Not Depend Upon
Heart with fittings (1)	Eye (2)
Brain with fittings (1)	Ear (2)
Nose with fittings (1)	Arm with hand/fingers (2) (2) (5)
Stomach with fittings (1)	Leg with foot/toes (2) (2) (5)
Liver with attachment (1)	Penis (1) or Vagina with Uterus (1) (1)
Lung with fittings (2)	
Kidney with fittings (2)	

At a glance we can see that the parts supplied in the greatest number—fingers and toes—are the ones we can lose with the least amount of hampered function, another verification of what we have already begun to suspect: in the human body, multiplicy seems almost to equal dispensable.

Let's consider the non essential parts list first. Almost all of these parts, which we can survive without however costly the loss, are supplied in duplicate. We can live without eyes—our design calls for two. Ears the same. Arms/hands, legs/feet—ditto. Duplication would seem to equate with non essential, the pattern being broken only when it comes to the penis or vagina with uterus, parts which are not supplied in duplicate even though individual survival is not dependent upon them. Species survival, however, *is* dependent upon them, a dependency that will be dealt with later.

If, as our overly generous supply of accessories would indicate, we have been produced by a profligate Workman or Work Force, why is our supply of non essential but eminently useful organs and appendages as limited as it is?

If we are allowed extra accessories—a greater number of teeth than the big cats have, a surplus of nails over claws, more hair than we need or want—why not extras when it comes to sensory organs?

Why should the human form be limited to two eyes? Surely an extra set of eyes, or even one extra eye, strategically located on the back of the head, would be of enormous survival value. And why not a second set of ears, possibly in the pelvic area or imbedded in our knees, to aid our hearing?

Is there some specific reason that the sensory organs duplicated in the human form are not supplied in triplicate or quadruplet?

And why only two legs? As any infant knows, it is a struggle to learn to balance on two wobbly extremities. Why was the prehensile tail, if there was one, eliminated rather than strengthened and allowed to evolve into a third, possibly retractible "leg" to help to support and balance us, a la the kangaroo?

Why were we shown so much generosity when it came to decorative accessories and so little generosity when it came to truly useful sensory organs and appendages?

Let's turn from the list of non essential parts to the list of essential parts.

First, we will surely note that when it comes to almost every organ essential to human survival we are given only one. No multiplicity here; even duplication is rare. *Why?* Why did the Creative Intelligence we are postulating decide, in the planning board stage, to become so cheap and niggardly when it came to the truly essential components of the human body?

Why only one heart?

Two lungs are provided. Why not a spare heart, with one heart placed symmetrically on each side, so that if one falters and fails the other can pick up the extra load and allow life to continue? With a Divine Being or Life Force so seemingly addicted to symmetry and duplication—two eyes symmetrically placed on either side of the face, two ears on either side of the head, two arms on either side of the upper trunk, one leg extending down on each side from the pelvic girdle—why is this scheme not carried through in the chest with one strong heart beating away on either side?

The heart is not the only essential organ supplied singly. Without brain function, life ceases, yet we have only one brain. One nose with fittings, without which we cannot breathe, or live; one stomach, one liver. This seems a perilous state of affairs: to make life dependent upon organs supplied singly. It should prompt us to take a closer look.

For the sake of argument, let's continue to assume that the human form is not in fact the product of a typically screwed up Committee, with every member tossing in her/his own pet idea, no matter how idiotic, and the result determined by majority vote; also that it is not the product of blind chance, a Force that didn't know what in hell or in heaven's name it was doing.

Assuming instead—admittedly an unproven assumption—that the human form was instead lovingly blueprinted, in every particular, by a supremely intelligent Being or Force, then we owe it to ourselves to look more closely to see if we can find anything approaching method in this apparent madness.

Why multiplicity of the most frivolous, duplication of the non essential, and only one of the essential?

A speculative answer:

Man looks around and sees everything in terms of duality: night and day, heat and cold, good and evil, body and spirit, male and female, life and death. And the most profoundly felt duality of all: the self and the non self, the creation and the Creator.

With two legs we stride upon our lovely earth. With two hands we do our work, with two arms hold our loved ones. Through two eyes we see, through two ears we hear.

Through duality we experience duality.

But is duality *reality?*

We know our senses lie to us. We can see that our earth is flat when in actuality it is a sphere. We can touch the solidity of the chair upon which we sit, the table upon which we write, yet we know that this too is an illusion. Is the duality we experience through our dualities also an illusion?

Let's look at the brain, one of the organs upon which human life depends.

On our list of parts, we have indicated that each body has only one brain. However, the largest and most human part of the brain, that part which evolved the most recently and governs voluntary movement and mental activity—the cerebrum or neocortex—has the appearance of a double organ consisting of two identical hemispheres joined together by nerve fibers.

Does this mean that, with our highly developed nervous systems, we have in actuality two brains rather than one?

The best answer for this is in terms of how the brain functions.

If one of the hemispheres of the brain is surgically removed early enough in life, the remaining hemisphere will so organize itself that all brain functions will be met and the individual will grow up without impairment. This ability on the part of the brain to adapt itself is soon lost, however. Once normal organization has occurred, the surgical removal of either hemisphere will result in impairment. Each hemisphere, while working cooperatively with the other, is a specialist, and it takes both hemispheres—full cooperation between two specialists—to cover the full range of activities governed by the brain. An apparent duality is, in respect to the way it functions, an operational unity.

Let's look at the heart.

The heart too has the appearance of a double organ. An interior wall divides it in half, each half consisting of a small upper chamber and a large lower chamber. If each half performed the same function, we would in effect have two hearts, just as we have two lungs and two kidneys. But the two halves of the heart do not duplicate each other's work. Here again an apparent duality is, in terms of functioning, an operational unity.

Regardless of the deceptive physical appearance, we have only one brain, upon which our life depends.

Regardless of the deceptive physical appearance, we have only one heart, upon which our life depends.

Is the duality we perceive all around us—in particular is the duality we *feel,* the *me* as separate and distinct from the *not me*— also a deception?

The human body is so constructed that it must be constantly replenished with food, water and air. To avoid dying of its own accumulated poisons, it must also throw off its waste products. For this intake and outgo, we have been supplied with several essential organs without which we could not survive.

With the primary sensory organs, and also with many of the body's glands, the Being Who or Force Which designed us—still assuming that One did—seemed to go along with the old cliche that to put all of any body's eggs in one basket is risky, but this careful, common sense approach was abandoned when it came to the more essential digestive and eliminative organs, There is only one mouth, one esophagus, one stomach, one liver.

The body is also dependent upon elimination of waste products for survival. With the exception of the kidneys, this work too is accomplished by organs supplied singly: that part of the large intestine known as the bowel/rectum/anus, the bladder, the uretha, the skin.

The following simple statement would seem to cover the facts:

The human body is dependent upon nourishment for survival, and nourishment comes through the one: one mouth, one esophagus, one stomach, one liver, one length of small intestine, one length of large intestine.

The human body is also dependent upon elimination of wastes—purification—for survival, and this purification is accomplished by the one: one bowel/rectum/anus, one bladder, one uretha, one skin.

Survival depends upon the one; life depends upon the one.

To summarize.

Bodily construction would seem to tell us this:

Multiplicity	= dispensable.
Duality	= the non essential.
Oneness	= the essential.
Oneness	= unity.
Unity	= life.

Before we can seriously argue this, however—argue that our bodies "prove" that oneness/unity is the underlying source of life, that oneness/unity alone sustains life while duality and multiplicy merely add interest and enjoyment—there are problem organs that must be dealt with:

1. The lungs essential to life yet supplied in duplicate.
2. The kidneys, ditto.
3. The penis/uterus with fittings, only one per body yet non essential to individual survival.

Can we make a case for these three as exceptions to a valid general rule?

Let's tackle the problem breathing apparatus—the lungs—by first considering the nose.

The nose, jutting out in the middle of the face, is divided by a partition into two nostrils. Each nostril performs the same function of drawing in and expelling breath. In this sense, we have two organs joined under a common casing. So why were the two organs not separated as the eyes and ears are separated, with one nose situated under each eye? With breath essential to life, this would seem a safer arrangement, also quite possibly a more attractive one, once we got used to it. With two separate noses, each containing one nostril, there would be a lessened chance of both nostrils being severely damaged and put out of commission with a single blow. If one nose were rendered inoperative, the second could take over, adjust its work load, and life could continue without interruption, as when one eye is blinded or one ear deafened. If I can see the advantages of such an arrangement, why wasn't it thought of back when in the drawing board stage of human life (still assuming there was such a stage)?

Possibly it was—and just as quickly discarded.

Possibly there was a reason behind this apparent slap-happy disregard of an easy safety precaution, an important clue as to purpose built into the human body.

While physical life depends on food, water and oxygen, the most compelling of these needs is for oxygen. The body can survive for weeks without food, for days without liquid, but without oxygen the brain is irreparably damaged within minutes and death occurs soon after. Without ceaseless breathing in and breathing out, without the never faltering functioning of the breathing apparatus, physical life ends very quickly.

The breath of life. Breath *is* life. How do we process the breath that is life?

We take in breath through our one nose, divide it immediately into two nostrils, then send it down one passageway, the trachea, into the lungs, where it is processed through myriad air sacs.

A speculative question:

Was our breathing apparatus—built to supply the most vital of our needs—constructed in this fashion to give us a blueprint as to how human beings interact, on the most basic level, with the Being Who or Force Which designed and created us?

A speculative answer:

Yes.

We breathe in life—come to Life—through the One (nose).

We immediately split this unity, via a sense organ (the nose), into Duality: the creation and the Creator. Life is, or is experienced as, a separation from the Creator.

Following this there is a rejoining—the creation becoming one again with the Creator—as we send the One Breath of Life down through the one trachea into the lungs.

In the lungs the trachea divides into two main bronchi—another Duality: the Self and the Non Self, which is how we experience life—before dividing into smaller and smaller passages ending in millions of tiny, balloonlike air sacs, the alveoli, which process the One (Breath/Life) into the Many, the myriad forms we see all around.

The lungs are an exception to the general rule because they are that part of the structure built to carry the primary blueprint of what life is all about: the One converted—for our pleasure or instruction—into the Many.

Problem organs 2 and 3 remain.

2

EXCEPTIONAL ORGANS

The Surplus Kidney and the Missing Penis/Uterus with Vagina

> Ye shall know the truth, and
> the truth shall make you mad.
> Aldous Huxley

Can there really be "exceptions that prove the rule"? This would seem an impossibility. Exceptions to a proposed rule would seem to render the rule invalid.

There is this consideration, however: exceptions to any rule always stand out.

If we dress all in white except for a bright red sash, the sash will immediately catch our eye. If we dye our hair orange but give it one green streak, the green will scream.

Is there something unique about the penis/uterus with vagina which might have caused our Designer—if we had One—to give it special treatment, possibly as a means to catch our eye and force our attention?

If we think about it, we might find something unique about the penis/uterus with vagina.

On our parts list, there is an either/or situation only in two instances:

1. In the assorted glands, with one assortment "His," another "Hers." Glands manufacture the hormones which bring about sexual differentiation.

2. In the sexual apparatus itself, the penis or the uterus with fittings.

Let's deal with the penis, which is more visible and flashier.

Survival of the individual male is not dependent upon possession of a functioning penis, or even possession of a non functioning one. But survival of the species *is* dependent upon the possession and functioning of male penises. In this sense—and in this sense only—the penis is an essential organ, just as are the heart, brain, et cetera, and all other parts supplied singly.

The penis as a procreative organ constitutes no exception to what our bodily construction seems to tell us: Oneness = the essential.

SPECULATIVE COMMENT:

Had the penis been blueprinted into the male body primarily as an organ of pleasure, it would not have been supplied singly. Had it been intended primarily as a source of sensation to add interest and enjoyment to life, the male body would have been supplied with two penises, not one.

As an organ of *procreation,* the penis fits into the scheme of bodily construction.

As an organ or *recreation,* it breaks the rules of bodily construction.

So what? In this day and age, when we are finally struggling free of the Victorian strait jacket, finally learning how to fully and happily indulge ourselves sexually, that's the last thing we want to hear, or, hearing, care about.

With the penis, there is not only the slight problem, dealt with above, as to its proper functioning, there is also the interesting aspect of its placement:

> *But Love has pitched his mansion in*
> *The place of excrement.*
>
> Yeats

Why?

Is there some purpose in this or did the Great Designer blueprinting our bodies—if such a Designer exists—simply lack all aesthetic taste?

Surely the penis could have been placed several inches higher, possibly around the region of the navel, and likewise that interesting uterine fitting, the vagina, thus removing them from "the place of excrement." Wouldn't this have confused us far less, and done away with much of the guilt all too often associated with "natural" sexual pleasures? With this placement, the penis could have been excused from all excretory function, and the male body supplied with the same simple urethal outlet as the female.

Why not?

Was there some compelling reason this wasn't done, some specific purpose in the placement we have?

In fear that we are beginning to tread on unwelcome ground again, let's drop this for now and return to the problem of the second exception.

Why two kidneys?

When kidney function is essential to life, the fact that we have two kidneys blows away the entire speculative scheme (oneness = the essential) if we can't come up with some plausible reason for the existence of kidney duality.

Let's deal with the male body.

All the blood in the body flows through the kidneys every few minutes. Waste products of living cells are removed, plus any excess water and acid. This waste fluid—urine—leaves each kidney through a tube, the ureter, and empties into the bladder, a storage sac. It is then expelled from the body though another tube, the uretha.

In the male body, the uretha is sheathed by the penis and is used to carry semen as well as urine. A sphincter muscle at the base of the bladder, under both voluntary and involuntary control, tightens as penile erection occurs, shutting off the flow of urine to keep the two fluids, semen and urine, from intermixing. At any given time the penis, enclosing its uretha, can be used to urinate or ejaculate, but it cannot do both simultaneously. A choice must be made.

This is a curious situation indeed. With no other organ of the body is this type of decision forced upon us.

We can choose to close our eyes or keep them open, but we do not have to decide whether we will use our eyes to see with or to hear with. The same holds true for our ears. Though we use our noses both to breathe and to smell, these are done concurrently; we don't have the power to decide that for the next couple of minutes we'll breathe, then after that we'll smell. We don't have to decide whether to have our hearts pump blood or switch to some other task, nor do we have any control over what function the brain will perform. Our digestive system does its own thing without any go ahead from us, as does most of the purifying apparatus. Only the penis, with its dual, unrelated functioning, forces a choice.

Fortunately for the continuation of human life, this choice between urination and ejaculation can be made moment by moment, with a constant switch back and forth. Males can choose to urinate at intervals all day, then ejaculate once or twice before falling asleep. Were this not the case, if the choice had to be made once and for all, irrevocably, at, say, twelve noon next Tuesday, life would soon dribble to an end—*regardless of the choice made.*

If the choice—we are still assuming no medical intervention—were for ejaculation, the male body would soon die of its own accumulated poisons.

If the choice were for urination, the species would soon die out for lack of procreation.

What an interesting spot to find ourselves in.

For the individual male to survive, he would have to choose urination over ejaculation.

For the species to survive, however, he would have to choose ejaculation over urination.

Why in God's name would the Great Designer—if there is One—put his favorite creation, Man, in such a curious situation?

Was He/She/It trying to tell us something?

Throughout much of recorded history, there has been a tendency to link sexual abstention—celibacy—with goodness or godliness.

In many major world religions—Hinduism, Buddhism, Taoism, Jainism, Roman Catholicism and the Eastern Orthodox Church—celibacy has a long, established tradition. It is seen as a discipline through which an individual can advance in spiritual growth, or as virtuous proof of devotion to God.

In the *Upanishads,* a collection of philosophical/psychological treatises composed by Hindu mystics between the 8th and 6th centuries BC, ascetic elimination of all personal desires is propounded as the way to achieve unity with supreme bliss.

Vardhamana Mahavira, founder of Jainism, lived as an ascetic and founded Jainism by organizing a celibate clergy and an order of nuns.

Gautama Buddha, in the 5th century BC, formulated "Five Moral Rules" for right living, the fifth of which was, "Let not one be unchaste." For Buddha the basic evil is *tanha,* a selfish desire, above all sexual desire.

In the *Tao Te Ching,* a small book setting forth the beliefs of Lao-tse, founder of Taoism, it is written that the individual should not "act from any personal motive," he should govern himself to become again "as a little child."

Jesus encouraged the celibate life for his disciples: "Let those acccept who can."

In the First Epistle to the Corinthians, St. Paul wrote: "I would that all men were even as I myself," that is, celibate.

St. Augustine believed that it would be "joy unspeakable" if men could somehow procreate without having to engage in sexual intercourse.

Even where the above named religions were not in vogue, the linking of celibacy with religious belief or spiritual progress was not unknown. The Greeks had their Orphic cult, in which the subjugation of the flesh was aimed for as a condition of release for the soul, and the Romans their highly honored Vestal Virgins, white clad, white veiled women who took vows of celibacy and who, if found guilty of having sexual relations, were beaten with rods and buried alive, for who but virgins were fit to tend to the city's sacred fire?

Purity—chastity—goodness—godliness: thoughout history the linkage exists.

If there is any validity to this linkage—to the depressing, unhappy notion that celibacy is good for us, or helps make us good—our bodies, if we were intelligently and creatively designed, should bear witness to it.

Do they?

To remain healthy our bodies must throw off waste material. Daily, or more or less often, our bodies eject solid matter. This matter, properly formed, is phallic in shape. To remain physically healthy, we must eject a never ending series of penile forms. We must daily, or twice daily, or almost daily, give up the "penis" within ourselves.

Oh, shit, is it possible we are being told to give up sex?

As though to make this unwelcome "lesson" even clearer, the organ of love—the penis or vagina—is located right next to the anus, through which the feces are excreted, thus establishing a very strong intimacy by association.

> *But Love has pitched his mansion in*
> *The place of excrement.*

This would seem to make the point so emphatically that only the adult consciousness, carefully not seeing what it doesn't want to see, could fail to catch it.

Very young children, infants and toddlers, easily make the connection. Many a young boy fights furiously against being toilet trained, against being forced to give up, to surrender, to have swept away down the toilet a precious part of his anatomy, the very part that makes him like his daddy. Despite Freud's contention that females suffer from penis envy and spend their lives feeling incomplete because not equipped with this organ, young girls—not having yet read Freud and therefore unaware of how they are supposed to feel—rarely object to bowel training, to giving up a daily penis. Surrendering this phallic object daily apparently does not deprive them of anything of much importance to them.

Further evidence of the close relationship between feces and the penis is offered by some female bodies. If a tampon—phallic shaped—is inserted in the vagina, a bowel movement—also phallic in shape—will automatically eject it, in many though apparently not all women. To eject the feces is automatically to "give up" forms inserted into the vagina, and we all know what organ is most commonly inserted.

No matter how many approaches lead to this same unpleasant, and quite possibly unwarranted conclusion, it remains something we don't want to hear, so let's drop it and return to our earlier question:

Why two kidneys?

During our habitation of the human bodies we find ourselves in, though we share membership in the same species, we all live—in a sense—on different levels. On one scale we can experience anything from robustly glowing good health to dismally weak, debilitating ill health. In like fashion, we can live anywhere between two extremes when it comes to our actions, that is, our lives in interraction with others.

The man who rapes and kills a child is attacking life from a slightly different perspective than the man who, at great risk to his own life, will plunge into a burning building to save a child. The man or woman who smuggles into the country illegal drugs, or deals in them, lining up young people to sell to other young people, lives life on a somewhat different level from the man or women who devotes his or her life to caring for and teaching handicapped children. In any of these cases, a life is being lived, but the lives do not necessarily equate. A man who rapes and kills is subtracting from the sum total of health and happiness in the world while the man who saves lives is adding to this total.

In a sense we can be said to live a double life: a physical life and a moral one.

The ancient yogis looked upon the human body and realized that it could be rather neatly split apart—metaphorically speaking—into different spheres of influence.

That portion of the body beginning at the waist and extending down through the two legs to the feet can justifiably be considered the physical portion of the body. The major digestive, eliminative and reproductive organs are located here, and the legs/feet connect us with the physical earth.

That portion of the body from the waist to the shoulders can justifiably be considered the emotional portion of the body. The heart, universally considered to be the site of feeling, and the lungs, which process the "breath of being," are located here, also our arms/hands extend from this area, and it is with our arms/hands that we reach out to others, either to give or to take.

From the shoulders up is the mental: the head housing the brain.

A further symbolic division was made of the face:

The lowest third, containing the mouth, which connects directly with the stomach/digestive system via the esophagus, was seen as the physical portion of the face.

The mid-section of the face, containing the nose, which connects directly through the trachea with the lungs, was seen as the emotional portion of the face.

The top third, containing the eyes, which connect directly through the optic nerve with the brain, was seen as the mental portion of the face.

With this helpful topography, we can attack the question of the extra, problem kidney.

Why two kidneys?

SPECULATIVE ANSWER:

The lower portion of the body represents the physical aspect of life and the sensory organ of this area, in the male body, is the penis. The penis is the organ of desire/yearning/lust—lust for sex, for money, for power. We can conquer this lust or be ruled by it. If we are ruled by it—if we choose lust for the physical over all other values—we are exercising the penis, possibly overexercising it, choosing the penis' ejaculatory function over its urinary function.

We are in effect rendering the kidneys relatively inoperative.

If this is not a good choice, how could our Designer—still assuming we have One—let us know that it isn't?

One way might be by furnishing us with an extra kidney, to alert us to the fact we lead a dual existence.

In being—existing—we lead a physical life.

In action—interaction with others—we lead what could be called a moral life.

The duality of our kidneys:

An active kidney promotes physical health.

An active kidney—rendering the penis relatively inactive—promotes moral well being.

RECAPITULATION:

If we are not the products of chance but were designed—certainly not yet a proven fact but nevertheless a distinct possibility—the Great Designer might easily have built clues as to our meaning and purpose into our structure.

To review the questions originally posed and offer speculative answers:

1. *Why is the human body supplied with two lungs, two kidneys, two legs, two arms, two eyes, two ears, two testicles or ovaries, yet only one heart, one brain, one nose, one stomach and one liver?*

Because life is dependent on oneness/unity.

2. *Why was the heart placed toward the left rather than toward the right or at a tidy, non prejudical dead center?*

We must breathe or we die, but one lung can collapse and quit and life will continue.

Without brain function we die, but one hemisphere of the cerebrum, the largest part of the brain, can be traumatized or surgically removed, and though some impairment will result, life will go on.

The heart, though in appearance something of a double organ too, functions as a tightly organized unity. The body can not lose half of its heart and live. If any part of the heart goes, life goes with it.

In this sense, the heart—every chamber of which is vital—is the most basic essential organ we have.

That part of the brain which governs the heart would become, by extension, the most basic, essential part of the brain.

The brain is composed (partially) of two hemispheres. The left hemisphere, which governs the right side of our bodies, is the verbal brain. The right hemisphere, in charge of the left side of the body, rules over our intuitive, feeling states.

If verbalization is *the* most basic and essential of our mental abilities, then the left brain should oversee our most basic organ, the heart. This would place the heart toward the right.

If verbalization and intuitive feeling are of equal importance, the heart should be dead center and under the rule of both hemispheres of the brain.

If our intuitions, our feelings, our dreams, are *the* most basic, essential part of us, of being human, then the right brain should oversee our most essential organ, the heart, and the heart should be placed toward the left.

Heart placement would be an extremely easy way for the Great Designer to clue us in as to what matters most.

3. *Why does the female body have a much more pronounced waist than the male body?*

The lower portion of the body, from the waist down, represents the physical; from the waist through the shoulders represents the emotional. If the two portions are drawing apart, if the emotional is becoming less based upon the physical, this should appear where the two parts converge, at the waist. This is the place to cut them off at the pass. The female body does this to much more pronounced degree than the male body, indicating that women are ruled more by their hearts and less by their groins than men. Observation of most cultures would bear this out.

4. *Why is the male body customarily far more hairy than the female body?*

Hair is quite intimately tied in with the sexual. At puberty hair begins to grow where it hasn't been before, in the genital area, under the arms, on the male chest. Those ruled more by their groins and less by their hearts would quite naturally grow more hair as a manifestation of where their ruling passions lie.

5. *Why was the body so formed that most of us enjoy sex*

> *But Love has pitched his mansion in*
> *The place of excrement.?*

As a teaser to bedevil us all.

6. *Why, in the male body, is one organ, the penis, called upon to function for the discharge of both urine and semen?*

This is a simple and easy way for the Great Designer—if there is One—to make the point that we are confronted by an important moral choice.

7. *Why are some female bodies so formed that a bowel movement will eject a tampon?*

Possibly as a reminder that the day will come that the vagina, for the sake of health, must give up the penis.

8. *Why are we so built that, to remain healthy, we must periodically eject solid matter in phallic form?*

God knows.

Two oddities of the human form:

Oddity number one: Though we have two kidneys, they do not come as a matched pair.

The right kidney is a little lower and smaller than the left.

The left kidney is higher, longer and narrower than the right.

What possible sense does this make?

SPECULATIVE SPECULATION:

Our moral life is ruled more by feelings than words.

The left kidney—the heart-side kidney, the kidney under the rule of the intuitive right brain—is a little larger and a little higher than its near-twin for this reason: it has a slightly larger job on a somewhat more elevated plane.

For this reason it is longer and narrower—as the road to heaven is said to be.

Oddity number two: the hemispheres of the brain cross over for their spheres of influence, the left ruling the right side of the body, the right the left. The nose alone does not share in the cross switch from left to right and right to left.

Why?

If any organ or organs are going to be exempted from the crossover, why not those sensory organs from which we learn so very much, the eyes or the ears? Why the nose?

SPECULATIVE SPECULATION:

The nose alone—nose, nostrils, trachea, lungs, air sacs—holds the key to the mystery of being.

That is, the nose alone has the straight of it.

The human body has more than form to guide us. The form has color. Possibly there are clues to be found in our coloring.

3

THE HUMAN BODY—IS IT COLOR CODED?

Can We Follow the Yellow Lit Bod
To the Wizard of Us?

> Somewhere, over the rainbow,
> skies are blue
> And the dreams that we dare to
> dream really do come true.
> "OVER THE RAINBOW"*
> Harold Arlen-E. Y. Harburg

Since long before Dorothy was told to follow the yellow bricks to the Emerald City of Oz, color has been used to give instruction and direction. The flow of traffic is ruled by color. Hospitals, businesses, recreation areas often have color charts embedded in their floors. Items which are purchased unassembled often use color tips to aid the new owners in their assemblage. This use of color permeates our lives.

If man with his limited intelligence has been bright enough to develop this use of color, what about the Being Who or Force Which created life? Isn't it possible that the Artist Who painted the created universe did not haphazardly splash color around but instead carefully color-charted the whole creation?

Let's assume for the sake of discussion that this did indeed happen, that colors were not thoughtlessly thrown at the universal canvas but were instead carefully thought out and assigned. The human body, an interesting part of the creation, a body which would in time take its place as undisputed kingpin of the earth, would have been included in this procedure and had its colors carefully assigned too.

With this assumption in mind, let's take a look at the coloring scheme in us, on us and around us, excluding from consideration, for the moment, that rarity, the albino, where normal color pigmentation is lacking.

Consider the eye.

The pupil of the eye, an expanding and contracting opening in the iris through which light passes onto the retina, like the early Ford car comes in one color only: black. The iris, in contrast, comes in a variety of colors, yet from a very limited range. Why—among those with normal coloring—are there no red irises, or orange, or yellow?

The exterior wrapping of the human body, the skin, comes in a variety of colors also, yet again from a limited range. Why is there no race of blue-skinned people, or purple, or green?

Man does not exist in a vacuum; human life exists against the backdrop of a big, beautiful, colorful world. If we make note of the permissible range of colors of various bodily organs, then relate this range to the appearance of color in the world around us, possibly we can detect the Great Designer's "color code" and learn to break it.

Of the three primary colors—red, yellow, blue—one dominates our physical world. The vast expanse of the sky, as seen from the surface of our earth during daylight hours, is blue. Close to three-quarters of the earth's surface is covered by water, and the oceans, seas, rivers and lakes of our planet home are predominately blue. This would seem to indicate that the Great Artistic Designer Who created our world—if there was One—was excessively fond of the color blue. So why aren't we—an important and unique creation—wrapped in a lovely blue skin?

Turquoise, sky blue, royal blue, violet, purple. Why are these not the races of man?

In contrast to the splashy, lavish use of blue in the physical world, when it comes to the human form, the appearance of blue is extremely niggardly. In the absence of artificial dyes or genetic abnormality, there are no blue skins, blue hair, blue teeth or blue nails. If our skin is fair enough, we can glimpse blue in the veins under our skins. Apart from this, blue appears only infrequently in Homo sapiens, and then just a tiny circle of it in the iris of certain eyes.

Only two of the primary colors seem permissible for irises: blue and yellow. Except for albinos, where, in the absence of normal color pigmentation, blood vessels give the appearance of pinkness or redness, eyes can be blue, blue-gray, gray, blue-green, green (blue/yellow), that darkened shade of yellow known as brown, hazel (a light golden brown), or black. The third primary color, red, is eliminated in all its infinite variety. The iris of the human eye is never naturally, healthily red, nor any shade thereof.

Why?

For the largest and most visible organ of the human body, the skin, there is a switch in the primary colors allowed. Red and yellow are in, blue is out.

Why?

Why are blue eyes allowed but not blue hair, blue nails, blue teeth or blue skins?

If we check out the role that the color blue plays in the universe around us, maybe we can stumble onto a speculative answer.

SPECULATION:

Man locates heaven up in the sky, in the vast calm blue.

If this is a proper placement, then peace—happiness—heaven are to be found in that which is blue.

Hair is sexy. If heaven is to be found in sex, hair should be blue, like the sky.

Nails and teeth are bodily weapons. If heaven is to be found in warfare, in assaulting others or defending oneself, the nails and/or teeth should be blue.

The skin is our bodily surface. If to stay on the surface—touching, being touched—will bring us to heaven, the skin should be blue.

If, however, heaven is to be found in none of the above but rather through an elevation of interest from the physical through the emotional to the mental, then the eyes—located on the mental level, connecting directly with the brain—should be the one bodily organ allowed to share the heavenly blue of the skies.

****

In the normally colored human body, red has been banished from the colors allowed in the human eye.

Let's check for red in the world around us, and in ourselves, to see if we can find a possible reason for this.

All life on earth is slavishly dependent upon the sun. While the sun is yellow/white all day in our sky—when one can see it through the clouds, haze, fog or smog—it is orange/red when it rises and red when it sets.

The fluid which courses through our bodies, bringing nourishment and growth, is red.

When cut, we spurt red.

The erect penis, engorged with blood, in the fair skinned body flushes pinkish/reddish.

Lips, both facial and genital in the female, are pink or reddish, as are nipples.

For us, as with the sun, the beginning of life (sexual activity) is reddish hued, and death, the end of it, if untimely and violent, if often red too.

But in normally pigmented bodies, there are no red eyes.

Why?

SPECULATIVE ANSWER:

We can live our lives primarily on a physical level, paying attention to the demands of the penis—to its lust for sex, for money and those material things that money will buy, for power over others—enlisting the rest of our bodies, our hands, arms, sensory apparatus, lungs, heart and brains, in penile servitude, expressing this attention to the penis by being satyrs or rapists, thieves or muggers, drug dealers or politicians.

If this doesn't do it for us, we can elevate our interest and live on an emotional level (waist to shoulders), listening to the yearnings of our hearts for a more intense life, putting all of our body's energies at the service of the heart and its enjoyment of those delicious excitations and delightful palpitations that come with falling in love, with the quickening of the pulse and the aching wish to hold and be held, to love and be loved, endlessly, endlessly in love.

Or if that doesn't quite do it, we can elevate our interest one more level, to the mental, and put our physical and emotional natures under the control of the mind.

The eyes exist on the mental level, connect directly to the brain, and should be colored accordingly, cluing us in to the profit or loss of the mental life.

Red is the color of violent death, the color that daily precedes the plunge of our world into darkness. In the direction of traffic, red signals STOP. All over the world, while green, an allowable color in the eye, means safety, red means danger.

If the elevation of interest from the physical through the emotional to the mental is dangerous, if it means a plunge into darkness, into death, then the appropriate color for the iris of the eye is red.

On the other hand, if such an elevation of interest, far from being dangerous, closes out, or tends to close out, danger and untimely, violent death, then red should never be employed as a color in the human eye.

****

What of green, that lovely color that *is* allowed in the eye?

As we look at our world, particularly during the marvelous, bursting days of spring, we can have little doubt that green is the color of renewal.

When the yellow of our sun blends with the blue heaven of our skies, rebirth results.

Yet green, very much like blue, can be found visibly in the human form only occasionally in certain irises.

The green of renewal, like the blue of heaven, is restricted to an infrequent appearance in the eye.

Yellow, in specks or in that darkened form we call brown, is also a permissible color in the eye.

Is this a plausible coloring?

It is through our yellow sun that we see. Without light we would all be sightless. Yellow would seem to be the color of seeing, of awareness, of knowing.

SPECULATIVE SUMMARY:

Although we can have yellow in darkened or speckled form in our eyes, no one of us housed in a human body sees well enough, clearly enough, to have pure yellow eyes.

If those of us in fair skinned bodes would find peace (blue), renewal (green) or awareness (yellow), we must go to the mental level, to our eyes.

****

Red means danger, everywhere in the world. It is the color of the spilling of blood and the color of the dying sun just before our world is plunged daily into darkness.

Apart from this, and the pinkish/reddish tint of our genitals, lips, nipples, and nails, where else do we see red?

When angry, we see red.

We can be red faced with shame, or red in the face and neck due to the angry red blood pounding through us. If anger has a color, that color is red.

Yet if we look at our bodies—at our genitals and those other bodily parts most apt to participate in sexual contacts—and assign a color to sex, that color would also have to be red.

If sex and anger share a color—as they do—is there any connection between the two?

Is anger a primary emotion in sex?

Is sex an expression of anger?

American slang would seem to answer these questions in the affirmative. The common interjection *Fuck you!* is not meant as a loving suggestion. To give someone the finger is both sexually suggestive and an open expression of contemptuous anger. *Go stick it—stick it up yours—up yours,* verbal equivalents of the gesture, do not express a friendly sexual interest. The word "attack" can mean assault in public or private war, with penile like weapons or with the penis itself.

Studies with monkeys have shown that anger is an essential ingredient in sex. Another study proved that the same was true for rats. A group of rats was placed in luxurious surroundings, with ample space, an overabundance of food, all the amenities for the good rat life. Mating became less frequent, then ceased altogether. Without frustration to spark anger, there was no sex drive.

Observation of non human animals in the wild has shown the close tie between assaultive behavior and sexual activity. A young male caught in a vulnerable position by an older, stronger male where flight is not possible will drop into the passive female sexual posture, allowing the stronger male to mount him. The older male will make a few thrusting moves, then will dismount and amble away, his aggressive mood defused.

A study among male college students showed the same underlying connection between assaultive and sexual behavior. Subjects were deliberately angered, then given—they were led to believe—a chance to retaliate. Prior to the opportunity for retaliation, most were shown erotic material in one of four subgroups: with high or low positive erotic content, or high or low negative erotic content.

"Results indicate that exposure to affectively positive erotic stimuli significantly reduced retaliatory behavior by angered males to a level below that exhibited by subjects exposed to the neutral stimuli and by those in the no-exposure control group.

"In contrast, relative to baseline controls, subjects' exposure to erotic stimuli that were reported to be disgusting and unpleasant slightly enhanced subsequent aggressive behavior."[1]

Anger can be channeled into sexual arousal; blockage of sexual arousal increases the level of anger.

SPECULATIVE SUMMARY:

When blood circulates through the body in its proper channels, it carries with it health and growth. When it bursts out of or leaks out of its proper channels and we bleed, either internally or externally, life is endangered.

When the pink/red natural to the genitals—the passion natural to the genitals—stays confined within its proper sphere, the physical, health and new life result.

When the pink/red of the genitals rises above its natural level, firing up the emotions, this intensifies the purifying work that must be accomplished, for the body to survive, by the heart and the lungs.

When the pink/red natural to the genitals—the passion natural to the genitals—rises even higher, past the emotional level to dominate the mental level, the head, and expresses itself on the face, it becomes a matter of anger and shame.

When red bursts forth onto levels where it doesn't belong, life is endangered. All signals scream STOP! Keep red in its proper place.

****

Unhappily, with both form and color, our bodies seem to be screaming at us, GIVE UP SEX! Or if not that, at the very least the message seems to be: With regard to sex, be careful—keep it in its proper place.

Why so much negative accent on an activity that is, after all, absolutely essential for the continuation of the human race?

SPECULATIVE ANSWER:

Not all forward movement, all progress, is completely benign, completely risk-free, as those of us living in the age of the atom bomb know only too well.

When man first tamed fire, this was an enormous leap forward. He could keep himself warm and comfortable—if he didn't burn himself up.

The invention of the wheel was another gigantic forward step—as long as the creature pulling the wheeled vehicle watched what he was about and didn't inadvertently initiate such quick motion he was run over by what he was pulling.

Ships took man aroving on the high seas—and oftentimes capsized and sank.

Airplanes went soaring into the skies—except for those that reversed direction and went plummeting into a diving crash.

The worldwide communications explosion of the last century has brought all of mankind much closer—so we can frustrate, anger, misunderstand and wage war against not only our near neighbors but our distant neighbors as well.

With nuclear fission we can envision an energy rich, drudgery free future for ourselves—if we don't sicken ourselves with radiation or blow ourselves up.

Progress brings risk.

If man is the result of an evolutionary process, then one of the gigantic forward steps he made was in the development of his feet.

His thumblike big toes moved in to where they were parallel with the other toes. In this position the big toes served to push-off and lengthen the stride.[2]

The heels came to rest on the ground to help bear and balance his weight.

He developed "shock absorbers," a long, springy arch in each foot.

These three changes are believed to be, by those who believe in the evolutionary process, the most significant changes in man's anatomical evolution.

If man did not evolve but was created substantially as he exists today, the One Who designed him gifted him with a unique foot.

Man's foot is "a masterpiece of engineering and a work of art." Leonardo da Vinci.

"Man's foot is all his own and it is unlike any other foot. It is the most distinct human part of the whole of his anatomical makeup. It is a human specialization; it is his hallmark, and so long as man has been man, it is by his feet that he will be known. . ." Frederic Wood Jones.

Man's unique feet "made him the only full time, erect-postured biped in all nature."[3] His hands, no longer needed to support body weight or to help with locomotion, were freed for countless sophisticated tasks. This made possible the accelerated development of his brain. It also changed coital postures, enabling man to indulge in positions unique in all nature.[4]

Man's upright posture brought with it a change in what provoked sexual stimulation. The role of odor "was taken over by visual excitations, which, in contrast to the intermittent olfactory stimuli, were able to maintain a permanent effect." Sigmund Freud.

Once man became upright, with his genitals fully in view, once sight rather than smell became the prime source of sexual stimulation, sex ceased to be periodic. It became instead—for the human animal only, a distinctly human experience—a year round, year in and year out, activity.

Fine and good, and all in the name of progress. But progress almost always brings risk. Are there risks inherent here too, in year round, year in and year out, sex?

Our bodies seem to tell us that yes, there are.

Why?

SPECULATIVE ANSWER:

In standing erect, man was pulling away from the physical, stretching up toward the sky. He was putting distance between his head, the mental, and the physical earth beneath him. Sex as procreation is a physical activity. If man, exposed to constant sexual stimulation by his new posture, allows this stimulation to rule him, if he, in effect, having drawn himself up from the physical, now drags the physical up after him into his head, allowing his emotional and mental life to be ruled by his sex drive, he has clearly, self evidently, reversed the very process his upright posture set in motion.

If sexual obsession exists, it overwhelms and buries the very spirit that caused man to straighten up to reach for the sky in the first place. It defeats the very spirit that makes him human, man the unique, the upright biped, Homo sapiens, the one who is wise because he thinks.

So how to deal with this risk, how to avoid mucking up the whole damn thing, the great and glorious upright surge?

Let's return to a consideration of skin color.

All normal skin color comes from four colors. The skin itself is white. A pigment in the skin adds a touch of yellow. A third color, black, is added through the presence of granules of melanin. Red is added by the blood circulating through the minute vessels of the skin.

White—yellow—black—red—mixed together give our bodies skin tones ranging from the very fair, almost pigmentless, to ebony black.

Differences in the mixture have brought about an informal classification of man into five races based upon skin coloration: white, yellow, red, brown and black.

Few things in man's history have been given the importance that skin color has. At the same time, skin color has been given next to no consideration. We have seen without seeing. Skin color has been right under our noses for centuries, with rarely or never a thought to the richness of meaning it might have.

Black. White. Let's deal with these first.

The sun, a red ball of fire, slips below the horizon daily, plunging our world into darkness.

Man is a diurnal animal. Historically he has labored by day and slept at night. During the dark he slips into that little death we know as sleep. The blackness of night becomes, for much of mankind—though this is not universally true—the color of death.

Historically the night has not only been the time for rest, but also for such recreation as is available, which includes that activity which has long been considered man's favorite indoor sport, sexual intercourse. Black becomes not only the color for death, but also one of the colors for sex.

This black association linking death and sex was expressed for centuries in the English language, when the words "to die" also meant to reach climax in sex.

Black = death, black = sex. Does black = death of sex?

While our lifegiving yellow sun dominates our days, moonlight refreshes our nights. The silver-white moon, shining solely with reflected sunlight, appears to be a source of light but is not.

In our world—if created by a Great Artistic Designer Who color coded it as we are still assuming—white becomes the color of illusion, of fraud or hypocrisy.

But brilliant sunlight, pouring down upon us from a yellow sun, is also white, the pure white light of the source of all life, so white in our world is the color of purity as well as of illusion and fraud.

If our bodies tell us—if the coloring of the world tells us—if some deep underlying instinct tells us, the same instinct that told us to crawl up from all fours, walk upright and reach for the sky—that sooner or later we must give up sex, or must at the least outgrow the all too human tendency to be dominated by the lusts of the penis, how can we attack this problem?

Two different approaches suggest themselves almost immediately, the two most common approaches used in attacking any addiction:

1. We can outgrow—outwit—the shrill insistence of the penis by turning a deaf ear, repressing our awareness of its demands, clamping a tight hold on ourselves much as a tourniquet cuts off blood supply. We can, with inflexible

determination, decide simply to be so self righteously pure that we will starve the damn thing, with its dangerous, never ending lusts, until it dies of neglect. If this is the course we decide to take, what is more fitting than that we should wrap ourselves in a lovely white skin?

We will take on the white mantle of purity—and, still under the sway of that which we are determined to eradicate, become the purest hypocrites under the sun.

But we know who the enemy is—the blackness of sex—and will of course angrily, enviously, attribute to a black skin a sexual prowess fearfully beyond us, and we will rarely or never question our superiority to such blackness, and our right to rule over it. We know that in opting for whiteness—for purity—we have taken the right course to rid ourselves of the dangerous black death.

This is one way to deal with the dangers presented by the lusts of the flesh.

Another way we might deal with them:

2. When faced with what seems an uncontrollable appetite, if we scorn the way of repression, knowing full well it leads to the stink of hypocrisy, we can decide to tackle it in exactly opposite fashion: feed, feed, feed the appetite hoping to reach, in fact to go well beyond, nauseatingly beyond, satiation, and in this fashion bring it to heel, an approach used successfully today in the cure of many addictions.

If this is the course we decide to take, what is more fitting than that we should wrap ourselves in the cloak of what we are openly, courageously challenging, that is, wrap ourselves in a lovely black skin?

Because this is a decision made on the mental level, however unconscious the process, naturally we must cover our heads with a kinky, woolly black hair, pubic-like hair, the sexiest of all sexy hair.

Broad, flaring nostrils will show the world—for those who have eyes to see—that there is a marked diversion in the ins and outs of our quest, that we are inhaling life in huge gulps of painful passion in order to expel it again.

Our mouths, though on a mental level, connect directly with the physical, so to acknowledge the course we are following we will give ourselves wide, deliciously full lips, evolving noticeably beyond the thin lip formation of the big apes and our hypocritical white skinned brothers.

Because the white-skinned search for purity through repression and denial infrequently meets with success, the hair of white skinned people is curly-wavy more often than not, pubic hair undergoing a frantic straightening process that rarely really takes. Noses are narrow, the ins and outs pinched together—breathe in repression, breathe out denial—and mouths share the thin lips of the big apes, the mouth of the human martyr, death before dishonor.

Two approaches to the same basic problem, how to avoid dragging up to the mental level—permanently—the physical world we are standing upright to free ourselves from.

In general, though in some instances this correlation breaks down, those with white skins have lived in the colder climes and those with black skins in the warmer.

Denial/repression is a colder phenomenon than indulgence in sexual feelings.

We are never likely to see a bumper sticker that reads, "Cold to trot." When we get worked up, we get heated up. Sex is warm, heading toward hot. Warm air rises, the penis rises. Denial pushes downward toward the cold.

Denial = white = cold.

Indulgence = black = hot.

Let those of us who think we can lick the sexual challenge through denial/repression head toward the north, toward the ice and snow.

Let those of us who think we can lick this problem through indulgence leading to aversion head toward the south, toward the heat of the equatorial sun.

Or possibly those who headed north were persuaded, through the coldness of their world, that denial was the way to go, while those who headed south were persuaded, through the heat of their world, that indulgence/satiation/aversion would work better.

It seems safe to say that, for the majority of mankind, neither of the above approaches has yet worked. As of this writing, both races stay mired—on the mental level—in the physical/sexual world, with the white, as is the way of the self righteous hypocrite, still giving endless "help" to the black, demeaning and exploiting him to prove to him the error of his ways.

This white-black "brotherly" relationship parallels quite closely, in possibly an informative way, two approaches to the breaking of alcohol addiction. One alcoholic, proud possibly to the point of unyielding arrogance, may attempt to quit cold turkey, without help, through denial, an attempt that may or may not work. Another, having seen or experienced the failure of the denial approach, may wisely and prudently seek help in a clinic or hospital, where he will quite likely be treated in some such fashion as this:

He will be given all the alcohol he could wish for, one drink after another after another, until he is too sick to want any more, then be forced to drink even more while placed in front of a mirror so that he is treated to the sight of himself vomiting. For this "help," he will pay quite handsomely—as the black race has always been forced to pay dearly for the "help" given him by the white.

Neither black nor white appears in the rainbow, and in the strictest sense neither is a color. White reflects—rejects—all light rays, black absorbs—accepts—all rays. As detailed above, white is a denial, black an indulgence. But what of those skins that have an actual color, skins that are neither black nor white?

Anthropologists today divide mankind not into the five races based on skin color current in my childhood, but into three main races based on various considerations. These three races are:

1. Caucasoid—wavy hair, narrow noses, white to dark skin.
2. Negroid—woolly hair, broad noses, brown to black skin.
3. Mongoloid—straight hair, flat faces, yellow, brown or red skin.

The white and black we have already considered.
Let's look at the yellow.

Our sun is yellow/white, yellow tinged, which makes yellow the color of seeing and awareness. But any human being who takes the time to be aware of his situation, who stops to think of himself as a microscopic bit of life on a small, insignificant planet spinning around on its axis, at the same time whirling in an elliptical path around a distant sun which is only one star in a galaxy of billions of stars, with the entire galaxy moving through space at a fantastic speed, and this only one galaxy in a universe of billions; that he comes into life howling with his first breath and from there proceeds inexorably toward a certain fate, the ultimate cessation of his breath, beyond which cessation he may or may not still exist—all of this is surely enough to give anyone pause. As Hamlet mused, "Thus conscience does make cowards of us all!"

Yellow = awareness.

Awareness makes us cowards.

Yellow makes us cowards.

Yellow assures a cautious, middleground approach, an approach which has been lauded, quite rightly no doubt, as wise, the inscrutable wisdom of the east.

If we're not sure we trust a given situation, if we feel uncertain and suspicious—above all, if we are of a cautious, yellow tinted nature—we look at our situation askance, that is, sideways, obliquely, not quite straight on. To express this cautious approach, we would naturally go to the eye. The epicanthus is a fold of skin extending from the eyelid over the inner canthus or corner of the eye, a fold of skin so common in members of the Mongoloid race that it is called a Mongolian fold.

Those of yellow skin look at life askance through what is commonly referred to, by other races, as a slanted eye.

Brown is yellow darkened, yellow enriched with black. The brown skin takes a less cautious approach, but remains allied with the middle ground.

A red tinted skin also expresses a less cautious approach, verging toward an open confrontation, the warrior who will stand firm and take on all enemies, on whatever terms are set, even if that enemy proves to be himself.

The Mongoloid race proclaims its awareness of the primary problem—the need to overcome the lusts of the penis—through black hair, but black hair that is fine and straight, sexless hair, a pronouncement that while there is an awareness of the challenge, the challenge has been met and overcome. Or will be met and overcome, given sufficient time. A cautious, "wise" approach.

There is no indication that the middle way has yet been successful any more than success has crowned the efforts of the white or the black. All races at this writing remain in the race, breathing, struggling, dying.

But though no race as a whole has yet succeeded in its quest, every generation brings innumerable individual breakthroughs in every race, beings in human bodies who have become truly upright, in every sense, men and women who keep the physical where it belongs, beneath their feet, their hearts, once awakened in love, beating strongly, steadily in their chests without the stimuli of constant new excitations, and their heads in the sky, their eyes calmly, easily on the stars. These men and women, regardless of the material circumstances of their lives, step forth into a bright, beautiful, rich new world that some instinct in each of us urges us to seek.

ODDITIES:

For over a thousand years one half of all Chinese women had their feet bound, beginning in early childhood, in order to break the bones and turn the foot back on itself, a procedure that rendered these women all but completely crippled. This was done in order to win the devotion of Chinese men, who were literally enchanted by the tiny, misshapen, three to four inch "lotus foot."

The "lotus foot," according to some authorities, caused an outpouring of erotic feeling on the part of Chinese men of such intensity that the thousand year love affair it inspired has never been equaled anywhere else on earth. Is there any possible explanation for this seemingly wild aberration?

SPECULATIVE COMMENT:

If, deep inside, all of us have the feeling that we have to, *must,* break our ties to the physical, to the earth, and know in our hearts and minds that the penis symbolizes, more than any other organ, this unhealthy tie that must be broken, some degree of repression is apt to occur in our erotic responses. It could scarcely be otherwise.

But what if sexual response was suddenly no longer tied in so closely with the physical? Or if we could somehow persuade ourselves that it isn't? How our feelings would soar, how erotic we could become! A woman with tiny, tiny feet—no matter that this is an artificial condition, deliberately visited upon her; we'll forget about that!—has little contact with the earth. Her tie to the physical has been reduced to almost nothing. Why, she is almost an angel ready to float away! Consequently feelings lavished upon her, erotic though they may be—and observers say they were the most intensely erotic feelings ever recorded—are not all that dangerous. This is *not* a devotion to the physical through sex, or so the Chinese male was apparently able to persuade himself.

ODDITY:

The word "banana" denotes a fruit. In slang, it is used to signify a comedian, particularly in a burlesque show. In the plural—to go bananas—means to spout nonsense, to go crazy.

SPECULATIVE COMMENT:

The banana is yellow but penile in shape. On the face of it, that is a comic or crazy combination. Who on earth has ever been able to be intelligently aware of anything with his penis?

ANOTHER ODDITY:

The albino, that rare individual lacking normal color pigmentation, has, as a result of the missing coloration, extremely sensitive, weak eyes, eyes so oversensitive to light that the eyelids are kept partially closed in protection.

SPECULATIVE COMMENT:

In those of us with normal color pigmentation, shouldn't the fact of color—in ourselves, in the great world around us—increase our vision accordingly?

4

THE LAWS OF HEREDITY

When It Comes to Our Genes,
Are We Like Peas in Their Pods?

> I will praise thee,
> for I am fearfully
> and wonderfully made.
> *Psalm 139-14*
> King James Version

In my desk dictionary (The Random House College Dictionary, Revised Edition, copyright (c) 1975), the word "heredity" is defined:

"*Biol.* 1. the transmission of genetic characters from parents to offspring, dependent upon the segregation and recombination of genes during meiosis and fertilization. 2. the genetic characters so transmitted."

In the same dictionary, "segregation" is defined: ". . . 3. *Genetics.* the separation of allelic genes in different gametes during meiosis, resulting in the separation of their characters in the progeny."

"Recombination" is defined: "*Genetics.* the formation of new combinations of genes either by crossing over at meiosis or by segregation."

"Allele" (adj. form "allelic") is defined: "*Genetics.* any of several forms of a gene, usually arising through mutation, that are responsible for hereditary variation."

"Mutation" is defined: ". . . 3. *Biol.* a sudden departure from the parent type, as when an individual differs from its parents in one or more heritable characteristics, caused by a change in a gene or a chromosome. b. an individual, species, or the like, resulting from such a departure."

"Meiosis" is defined: "*Biol.* the maturation process of gametes, consisting of chromosome conjugation and two cell divisions, in the course of which the diploid chromosome number becomes reduced to the haploid."

"Gamete": "*Biol.* a mature sexual reproductive cell, as a sperm or egg, that unites with another cell to form a new organism."

"Diploid": "1. double, twofold. 2. *Biol.* having two similar complements of chromosomes."

"Haploid": "1. single, simple. 2. *Biol.* pertaining to a single set of chromosomes."

"Chromosome": "*Genetics.* any of several threadlike bodies, consisting of chromatin, that are found in a cell nucleus, and carry the genes in a linear order."

"Gene": "*Genetics.* the unit of heredity transmitted in the chromosome that, partially through interaction with other genes, controls the development of hereditary character."

There we have it—heredity. If we wonder why we have Dad's big nose but not his broad, charming smile, why we have Mom's stubby fingers but not her long, shapely legs, why we have Grandma's weak chin but not her piercing blue eyes, Grandpa's sloping shoulders but not his beautifully shaped head, we have only to remember genetics, the science of heredity, to have it all come clear.

Every human being inherits a set of genes from each parent. Cells in the human body are diploid; that is, every cell contains a double set of chromosomes. When the body reaches sexual maturity, gametes—sperm or eggs—are formed. During the gamete maturation process, reduction occurs and the diploid chromosome number becomes haploid, so that each gamete is haploid, that is, it contains only one set of genes-bearing chromosomes. When the egg and sperm unite, each tosses in a single set of twenty-three chromosomes to the new organism, bringing the total back up to the standard diploid number of forty-six.[1]

So there we are, inheriting half of our father's complement of genes and half of our mother's. But why not the half of his we'd love to have instead of the half we love him in spite of? Who decides, or how is it decided, which particular genes we're going to be blessed or cursed with?

What is "crossing over"?

Is my musical talent—an unheard of talent in my family—the result of a happy mutation, and if so, what causes mutations?

Why do I look as I do?

Why do I have the talents I have?

How did I happen to become me, with the particular combination of facial features, body configuration and personality traits that I have?

Pioneer work in the field of genetics was done by an Austrian monk, Gregor Johann Mendel (1822-1884).

Mendel worked with pea plants that bred true to type. Pea plants that bore yellow seeds always produced other yellow-seeded plants, and green-seeded green-seeded. He asked himself what kind of offspring would result from artificial cross-fertilization of true breeding plants differing in a single characteristic, and set about cross fertilizing them to find out.[2]

If a *tall* variety of true breeding pea plant was fertilized with pollen from a true breeding *short* pea plant, what height offspring would result?

Several different results seemed possible:

All the offspring might be intermediate in height between the two parent strains.

All might be tall.

All might be short.

Some might be tall and some short.

Actual result: all offspring plants were tall.

Mendel used the word "dominant" to refer to the characteristic which appeared in the offspring—in the above case, *tallness*—and the word "recessive" to refer to the characteristic which disappeared—in this case, *shortness.*[3]

He studied a variety of pea plant traits, cross fertilizing true breeding pea plants that differed in these traits:

Seed shape	smooth crossed with wrinkled
Cotyledon color	green crossed with yellow
Seed-coat color	gray crossed with white
Pod shape	unconstricted crossed with constricted
Pod position	axial crossed with terminal
Stem length	tall crossed with dwarf[4]

The results of this cross breeding led to the formulation of the "Law of Dominance." When plants showing contrasting forms of a trait are crossed, all of the offspring show one of the contrasting forms of the trait; the other form of the trait disappears.

It is not yet known how this dominance is achieved, that is, how one tall gene in a hybrid is as potent in achieving tallness in the pea plant as the two tall genes of a true breeding plant. To further mystify the matter, the fact that a given trait is dominant over its allele in one species does not necessarily mean it will be dominant in another species. "Black is dominant over white in rabbits and guinea pigs," but "white is dominant over black in the Leghorn chicken."[5] To further confuse the matter, in some cases neither characteristic of a pair of traits can dominate the other and a blending of traits results. This is known as *"incomplete dominance* or *blending inheritance."*[6]

Examples of complete and incomplete dominance in man: "dark hair color is completely dominant over light hair color, and dark eye color is completely dominant over light eye color. But skin color shows incomplete dominance," so that the offspring of black/white parents will show a "blending inheritance" in skin coloration.[7]

After observing the dominance of certain traits in pea plants, Mendel was curious as to what had happened to the recessive traits that had disappeared. He had the offspring (hybrid) generation self fertilize—and the recessive trait reappeared in the new generation, in unaltered form.

The ratio of the dominant to the recessive, in experiment after experiment, approximated 3:1.

This led Mendel to formulate the "Law of Segregation": The recessive trait that is hidden in the hybrid generation is not lost, since it reappears in subsequent generations.[8]

Further experimentation led Mendel to formulate the "Law of Independent Assortment": Every characteristic is inherited independently of every other characteristic.[9]

The above "Mendelian Laws of Heredity" are today considered principles rather than laws as it has since been found that there are as many exceptions to each rule as there are cases that follow the rule.[10]

Others besides Mendel began to develop "laws" of heredity, and in 1900, *"genetics,* the new science of heredity was born."[11]

A great deal of work was done with the tiny, red-eyed fruit fly, *Drosophilia,* "ideal material for. . . experimental work. In fact, it can be stated safely that in the early part of the twentieth century, more was learned about heredity from the study of *Drosophilia* than from all other sources put together."[12]

Pea plants—fruit flies—man.

****

To determine how heredity works in man poses a somewhat more difficult challenge than to figure out how it works in pea plants or fruit flies. An experimenter can cross breed pea plants at will, as Mendel did. Fruit flies obligingly reproduce at a rate of twenty-five generations per year and hundreds live comfortably in a milk bottle.[13]

It is not possible at present to breed people at will, to suit the experimenter, and the twenty-plus year gap between generations would make it difficult to follow the appearance/disappearance of given traits through several generations even if breeding were under laboratory control. Extrapolation from pea plants, fruit flies and other sources, plus thousands of years of observation of the human condition—"Johnny is certainly the spitting image of his Dad!"—have led to the belief that man is, in general, subject to the same principles or laws of heredity that have been proven to be operative throughout much of nature.

How valid is this extrapolation from pea plant and fruit fly to man?

The relatively young science of genetics has formed close ties to several other disciplines. From Mendel's day, it has leaned heavily on mathematics. "Today, statistics, probability and mathematical analysis are constantly playing a part."[14] Genetics has also formed a close alliance with *cytology,* the study of cell structure, has called upon physics, with the use of "x-rays, ultraviolet light, atomic energy, and other forms of radiation," and recently has formed a partnership with chemistry. "Under the title of *Biochemical Genetics,* this partnership seeks to understand the effects of heredity on the chemical reactions which are carried on in the bodies of living things. Or, to put it another way, this partnership seeks to explain how genes control the fundamental chemistry of life itself."[15]

Is it possible that it would be equally valid and rewarding—or even more valid and rewarding—to go at this the other way around, that is, to seek instead to explain how the chemistry of life—the chemical reactions which are carried on in the bodies of living things—*exert control over the genes?*

If genes make the man—as geneticists tell us they do—how many genes does it take?

No one knows. There is not even a reliable way to estimate the number. It can be assumed, rather arbitrarily, that "the number of structural genes lies somewhere in the range from 10,000 to 100,000," while no attempt is even made to "divine the number of nonstructural genes."[16]

According to current thinking, the genetic makeup inherited by any given individual is largely a matter of blind chance. If a brown eyed parent happens to carry a recessive gene for blue eyes, it is a toss of the die as to which gene—the dominant one for brown eyes or the recessive one for blue—will turn up in his offsprings' genetic pot. The same goes for a vast assortment of other traits believed to be inherited. Blind chance is further modified by more blind chance, and again by more chance. All in all, the entire hereditary package is extremely chancy.

Mendel's original "law" that each characteristic is inherited independently of every other characteristic has failed to stand up upon further experimentation. It has been found, instead, that traits are often inherited together, due to a phenomenon known as *linkage.*

While organisms may have hundreds of thousands of heritable traits, the number of chromosomes bearing the genes is limited. Therefore each chromosome must carry many genes. The genes on one chromosome, linked together, traveling together, are inherited as a package—unless, by chance, crossing over occurs.

So what is *crossing over* and why does it occur?

It is a breakdown in linkage and who knows? Current thought attributes it to chance.

The name expresses what happens, but to clarify further:

Line up twelve men on one side of a room, twelve women on the other. Have each line walk toward the other until the two lines meet, then have each woman begin to dance with the man opposite her. After a few steps, they stop dancing and part again. Each line heads back toward its own side of the room—except that one man crosses over, joins the line of women and walks back to the women's side of the room while one woman crosses over, joins the men's line and walks over to the men's side.

In essence this is what happens in the crossing over of genes.

The paired chromosomes of a diploid cell—a cell containing two sets of genes—duplicate themselves, and the newly formed *chromatid* ("*Biol.* one of two identical chromosomal strands into which a chromosome splits before cell division") come together and twine about each other, in what is known as *synapsis.* During this entwining, crossing over may, by chance, occur. When the chromosomal strands pull apart again, genes from strand A may possibly have crossed over and joined strand B, while B genes have joined strand A.

The effect of crossing over is that even genes that are "linked" together in a single chromosome "do not necessarily remain together from one generation to another. . ."[17]

What about gene changes, ie, mutations?

Genes (if they actually exist; modern geneticists do not all accept the idea that they do) "are extremely stable molecules. Therefore they do not often change."[18] However, mutation does occur. Genes, if they exist, are subject to change through chemical means, also through the use of x-rays: "radium emanations, neutrons, ultraviolet light, and other forms of irradiation were also effective, as were extremely high temperatures applied at the proper time in the life cycle."[19] H. J. Muller, Nobel Prize winning geneticist for his work on mutations, found that many of the mutations which he produced artificially were similar to those which occur naturally."[20]

But what causes mutations to "occur naturally"?

There is a wide variation in the mutation rate for different genes. Why?

Various other changes and disasters can occur to chromosomes on their way to supplying you with your genetic inheritance.

"Sometimes a piece of chromosome breaks off during gametic formation and gets lost. This is known as *deletion.*"[21] If the lost piece attaches itself to another chromosome, this is *translocation.* Heredity is altered as some gametes will be *deficient*—missing the chromosomal piece—while other gametes will have *duplicates*—they got stuck with what was lost.

"Sometimes two chromosomes which are not members of the same pair exchange pieces. This condition is known as *reciprocal translocation* or *segmental interchange.*"[22] This is not the same thing as crossing over, "which involves two chromosomes of the *same* pair."[23]

Inversion can occur—pieces of chromosome get turned upside down.

Also *nondisjunction*—"a pair of chromosomes fails to separate during reduction division,"[24] resulting in *heteroploidy ("Biol.* having or designating a chromosome number which is either more or less (usually by one) than the somatic number of the species.")

As we become acquainted with all the disasters that can befall our genes—all the possible ills the poor things are heir to—it begins to seem somewhat miraculous that any of us comes out in a more or less understandable shape, resembling, to some extent at least, our parents and our siblings.

Why do I look as I do?

Why do I have the talents I have?

How did I happen to become me, with the particular combination of facial features, bodily configuration and personality traits that I have?

Genetics tells us that this is how heredity works:

Inherited traits are determined by genes (if there are such things), and chance determines which genes we will get. Chance also determines how battered and bruised—properly placed, misplaced or lost, upside down or rightside up—these genes will be.

After segregation—the separation of allelic genes—recombination may occur through crossing over; mutations are possible; deletion or translocation can happen, reciprocal translocation or segmental interchange; or inversion; or nondisjunction leading to heteroploidy.

Let's apply this to imaginary case histories to see how all of this works.

Height is considered to be an hereditary trait, influenced/modified by environmental factors. So:

Six foot, two inch Harry marries five foot, two inch Mary and they produce three sons.

The oldest son attains an adult height of six feet, five inches, three inches taller than his father. Genetics tells us how this happened: it happened by chance.

The middle son grows up to be six feet, two inches, the same height as his father. Genetics tells us how this happened: it happened by chance.

Their youngest son attains a height of only five feet, ten inches, the same height as his mother's tallest brother. Genetics tells us how this happened: it happened by chance.

Another example: Eye color, hair color, and facial features are considered heritable traits, though they too can be influenced/modified by environmental factors.

Handsome Hans marries a plain Jane and they produce three children, two boys and a girl.

Their first born, a boy, "takes after" his father and is extremely good looking. Genetics tells us how this happened: it happened by chance.

The younger son takes after his mother, not his father, and is considered to have greater inner beauty than outer. Genetics tells us how this happened: it happened by chance.

The youngest, a girl, has such a mix of features that she doesn't look very much like either parent; she is thought rather unattractive as a child but in adolescence unexpectedly blossoms into a genuine beauty. Genetics tells us how this happened: it happened by chance.

Thanks to the genetic scientists, we now understand how we got to be the way we are.

Whether, with our green eyes and auburn hair, we most resemble Dad's brother Tom or Mom's cousin Beatrice, if Dad supplied a human sperm and Mom a human egg, the chances are close to one hundred percent that if the fertilized egg survives it will mature into a recognizably human body, with the standard operating equipment. Fertilized human eggs grow into human babies, not into baby chimps or orangutans. This is assured by genetic inheritance and, in distinction to superficial traits like stature, coloration or facial features, there is very little that is chancy about it.

****

There has been a great deal written in recent years about DNA (deoxyribonucleic acid), RNA (ribonucleic acid), and the cracking of the genetic code which has resulted in gene splicing and the creation of new life forms in the laboratory.

DNA, usually found in the nucleus of the cell, is seen as the master builder, in charge of the blueprints, while RNA, produced by DNA in the nucleus, forges out into the *cytoplasm* (*"Biol.* the protoplasm of a cell exclusive of the nucleus." *Protoplasm, "Biol.* . . the living matter of all cells and tissues") carrying "instructions" from the boss DNA as to what kinds of proteins are to be built.

The DNA boss' blueprint instructions are delivered by the "messenger" RNA to the ribosomes (*"Biol.* any of several minute particles composed of protein and RNA and found in the cytoplasm of a cell) where the protein synthesis occurs, that is, where the actual building takes place. The RNA in the ribosomes does not seem to migrate, but other, very short-chain RNA molecules have been found in the cytoplasm which can pick up amino acid (protein) molecules and move them about to wherever they are needed. Building proceeds apace and growth results, genetically determined.

If the DNA master builder blueprint calls for an *E coli bacteria* cell to be built, that's what is built—unless laboratory experimenters intervene and alter the genetic code, creating a blueprint that produces a new life form.

This gene splicing/creation of man-made forms is now being done on the bacterial level, with enormous implications for all forms of life.

Is man as vulnerable as the *E coli bacteria* to forces external to himself manipulating his genes?

Is he as vulnerable to the operation of chance as he would appear to be in his genetic inheritance?

To what extent, if at all, is man at the mercy of his genes?

Mendel could work out "laws" of heredity by experimenting with pea plants, but can these same "laws" be applied to man?

The tiny red-eyed fruit fly, *Drosophila,* can be bred generation after generation in the laboratory and a great deal learned about its hereditary patterns, but do these findings have any applicability when it comes to human beings?

The genetic scientists have answered that yes, heredity in man *does* follow much the same pattern as that observed in pea plants, *Drosophila,* and in numerous other animals and plants. "Everything we find in the inheritance of plants and animals is also found in human inheritance."[25] There is evidence in human heredity of "dominance, blending, segregation, recombination, independent assortment of traits, linkage of traits, multiple allels, multiple-factor inheritance, sex linkage, mutations, etc."[26]

In the absence of direct laboratory experiments, how has this evidence been gathered? Largely through a statistical attack, information drawn from family pedigrees, medical records, prison records, from the study of twins and random populations, et cetera.[27] All of these studies have led to the conclusion that man, like pea plants or red eyed fruit flies, is the product of his genes and the interaction of his genes with the environment.

Is this a valid conclusion?

Has the principle of dominance, formulated as a law by Mendel with regard to pea plants, really been "found in human inheritance"?

Two brown eyed people marry and have one brown eyed child and one blue eyed child. Both parents MUST have carried the genes for blue eyes, according to genetic theory, or they could not have produced a blue eyed child. Because the parents have brown rather than blue eyes—though they carry the genes for both—the genes for brown eyes must be dominant, and for blue eyes recessive. The grandparent generation is checked and, sure enough, one of each set of grandparents is blue eyed, the source of the parents' blue eyed genes. Through inference—that the parents carried genes for blue eyes, inherited in each case from a blue eyed parent—we have "found" the operation of dominance in human beings.

Oh, really?

Well, that's how it's been proved to work in pea plants and fruit flies.

But lots of things work differently for pea plants and fruit flies from the way they work for man.

For example, consider this series of statements:

Pea plants have not been endowed through genetic inheritance with the wings needed for aerial flight; therefore pea plants will never fly.

Man has not been endowed through genetic inheritance with the wings needed for aerial flight; therefore man will never fly.

Fruit flies have not been endowed through genetic inheritance with the large, powerful, ocean-adapted body of the whale; therefore fruit flies will never roam the high seas.

Man has not been endowed through genetic inheritance with the large, powerful, ocean-adapted body of the whale; therefore man will never roam the high seas.

Pigs have not been endowed through genetic inheritance with a body structure built for great speed; therefore pigs will never move across a land surface at a faster speed than a horse or a deer.

Man has not been endowed through genetic inheritance with a body structure built for great speed; therefore man will never move across a land surface at a faster speed than a horse or a deer.

Clearly man has accomplished things undreamed of, as far as we know, by the pea plant or fruit fly, and for this reason: either through genetic inheritance or through some other means, he has been endowed with a trait not shared with other life on this planet, or not shared in equal measure, and that trait is, of course, creative intelligence.

Creative intelligence has freed man from many of the limitations imposed by his genetic inheritance.

Due to his creative intelligence, man is not completely at the mercy of his genes.

Let's risk taking this one step farther and suggest this hypothesis:

HYPOTHESIS:

Man is in no way at the mercy of his genes.

Segregation—dominance—blending—crossing over—mutation—deletion/translocation—segmental interchange—inversion—nondisjunction. Let's eliminate the element of chance from all this inferred genetic behavior, put a creative finger on the trigger, and suggest the above hypothesis not in negative phraseology but in clear and positive terms:

HYPOTHESIS:

Man has free will, and *this free will extends even into his genes.*

Unlike the pea plant and the red eyed fruit fly (or possibly not unlike them), *man creates his own body to suit his own needs.*

Let's call this process "envisionment" and deal with it in the chapter to follow.

5

GENETIC VERSUS ENVISIONMENT

If Junior is "The Spitting Image" of Dad, Is He Sincere—Or Sincerely Repressed?

> Example has more followers than reason. We unconsciously imitate what pleases us, and approximate to the characters we most admire.
> Christian Nestell Bovee

In my desk dictionary (Random House College), the word "genetics" is defined: "1. *Biol.* the science of heredity, dealing with resemblances and differences of related organisms resulting from the interaction of their genes and the environment."

Back when I was young, genes and the environment were on different sides of an ongoing war: it was heredity *versus* environment. (*Environment:* "1. the aggregate of surrounding things, conditions, or influences.")

Judging from the above definition, this war has now been settled, and settled in a way that many wars are, by one combatant swarming over and declaring sovereignty over the territory of the other. "The science of heredity" is now defined as an *interaction* between the genes (the transmitters of heritable traits) and the environment, ie, genetics has now become an all encompassing package, heredity *interacting* with environment, heredity *plus* environment. The war is over—on the dictionary level at least—and heredity, having absorbed the opposition, is clearly the winner.

Has the enemy territory—the environment—really been conquered, made an obedient subject state to *genetics,* the "science of heredity"? Or is the new young science overreaching itself, proclaiming victories that so far exist on paper—the dictionary pages—only?

On the face of it, the word *genetics* would seem to pertain to *genes,* not to such environmental conditions as level of temperature, exposure or non exposure to light, proper nutrition, et cetera. But in practical application, it was found to be difficult if not impossible to unravel the role of the genes from the influence of the environment. "The two must work hand in hand."[1] Improper nutrition can cause stunted growth regardless of genetic inheritance; "a fault in a gene or a fault in the environment may each independently give rise to the SAME fault in development."[2] As Dr. George Corner put it, "The same malformation may result from *a bad egg in a good environment as from a good egg in a bad environment.*"[3]

But—if genes are this dependent upon the environment, if an unfavorable environment can cause malformation and good genes are no guarantee against faulty development, do we inherit anything at all worth considering through our genes?

Yes, say the genetic scientists. Although, due to the crucial importance of environmental factors, we can not be said to inherit characteristics or traits, we do inherit *capacities.*

Capacity: ". . . 4, actual or potential ability to do something."

Whether the *capacity* becomes actual or remains a potential depends upon the environment.

On the dictionary page, genetics, "the science of heredity," has taken over the enemy territory, the environment. But when we look at how this developed we find that the exact opposite actually occurred: heredity declared itself subject to environment, the genes being stripped of power to where they are no longer believed to pass on characteristics or traits, but simply an "ability to react."[4]

To react to what?

To the environment, of course.

Through our genes we inherit *potential.* How restricted or unrestricted is this potential?

In interacting with our environment, how much choice, if any, do we have as to what potential will be actualized?

Those who believe in conventional genetic theory offer us a narrowly restricted range of potential.

For example: It is believed by geneticists that hair, eye and skin coloring are all inherited. In the human eye, brown is believed to be dominant over blue. This means, according to conventional theory, that if a gene for brown eyes is passed along, tossed into the genetic pool, the offspring will have brown eyes—provided the environment is favorable enough to allow for normal development. The environment can thwart normal development or it can allow normal development to occur; nothing more. Its role is strictly that of a possible spoiler. The power of determination, in a "normal" environment, still resides in the gene.

Admit the influence of the environment and the new organism is still seen as having no power of choice. Instead, in conventional theory, it is seen as under the control of two factors rather than one, as subject to the mastery of its genes and also the vagaries of its environment. At no point—in conventional theory—is it believed that the organism can take a stand and decide for itself: "Hey, kid, we've got genes here—blueprint plans here—for brown eyes or green or blue, so—what do you say—let's have blue eyes, okay?"

This kind of fairly unrestricted potential is ruled out by genetics, the science of heredity.

What is the immediate environment of the human gene?

DNA (deoxyribonucleic acid), the transmitter of hereditary characteristics, has been shown to be the master builder, the brains of the outfit. DNA is found in each of the cells of a human body—in fact, exists there in an embarassingly large amount.

The immediate environment of human DNA is the cell it finds itself in, and the cell is a tiny part of a complex, magnificently organized structure with an intricate network of nerves, a heart to pump life-sustaining fluid though it, and a brain capable of storing a vast amount of knowledge, a brain that gives rise—though no one knows exactly how—to a mind capable of all kinds of wonderful things: thought, feeling, logical analysis, day dreams, night dreams, and the creative imagination that has allowed its owner to burst the restrictive bonds of its own genetic inheritance, to soar through the air, roam the high seas, speed across the land, in effect to master its environment in almost every way. Yet this powerhouse of a body, with its attendant magical ("mysteriously enchanting; not yet explained") mind, is seen as all but completely at the mercy of the DNA in the nucleus of its cells.

Recent research has shown that DNA does not work the way it was originally believed to work.

In the *Escherichia coli* bacteria, where DNA was first observed at work, biologists watched as the DNA duplicated itself segment by segment in the RNA, then the messenger RNA left the nucleus to carry the genetic message out to the cellular machine where the genetic command was executed.

It was all very neat and satisfying, and biologists assumed that this was how it worked in all living things.

Subsequent study, however, brought about a revision of thought as it was found that DNA in the more highly organized forms of life did not work in the same tidy, non troublesome way that had been observed in the *E. coli* bacteria.

In more highly organized forms of life, DNA did not duplicate itself segment by segment in the messenger RNA. Instead DNA was found to contain numerous unused strands, which the biologists promptly labeled "genetic gibberish." The messenger RNA matched up only with "sensible segments," ignoring the "nonsense interruptions," ie, the "genetic gibberish," before departing the nucleus. These so-called "split" genes—DNA split between *exons,* which are sensible, and *introns*, which are not—have now been found everywhere, in every species examined, with few genes not sharing this fate. "Some genes are spectacularly splintered, the champion to date being the collagen gene, which codes for the main structural protein of bone. It is broken into some 50 parts."[5]

Ah, how lovely. We have been looking for potential in human inheritance, and now the biologists report all this unused DNA, all this non actualized, *potential* genetic material.

So inured to the possibility of choice in heritable traits are the genetic research scientists that this unused DNA—the unused potential—is seen as "nonsense interuptions," as "gibberish." The exons—that part of the DNA which is duplicated by the messenger RNA and is expressed, thence the name—are the "sensible segments." This view is held despite the discovery that the exons are "usually much shorter than the nonsense interruptions," the exons averaging "around 100 to 300 genetic letters while introns average 1,000. Some introns have as many as 10,000 letters."[6]

Now this is curious situation indeed. The human cell has "maybe a thousand times more DNA" than a bacterium[7]—with no known explanation as to what all this extra DNA is for—yet when it is found that only a small part of the DNA is expressed, that exons are "usually much shorter" than intragenes, the so-called nonsense segments, no biologists suggest that possibly this has something to do with the wide range of human potential, or with the working of the wide ranging human imagination.

If all this unused DNA is really waste material—gibberish—containing nothing but nonsense—why do we humans have so much of it and the *E. coli* bacteria none at all? The human body is usually found at or near the top of the heap when it comes to the good, usable things in life, not loaded down with a lot of junk stuff. Why does each of our cells have "maybe a thousand times more DNA" than a bacterium has? Is all this DNA really "extra"?

Let's consider two parallel situations.

An experienced builder of mousetraps sets out to build one. He makes a list of the materials he needs, gathers this material, and builds his mousetrap. He winds up with a well built mousetrap and no excess material whatsoever, all neat and tidy in imitation of the way DNA works in the *E. coli* bacteria.

A second builder plans to build a cathedral but he is unfamiliar with the site upon which he will erect his structure. He doesn't know beforehand whether he wishes to build it with brick, stone, marble, wood, stucco or what, and won't know until he is actually on the building site. However, once he is on the site it is too late to order additional materials, so in making out his list, ordering and gathering his materials, he includes every possible building material along with the bonding agents needed to work with each, ie, nails, mortar, or whatever, also all the tools required for each and every type of construction. All of this adds up to an enormous amount of material, most of which he knows going in won't be used.

He arrives at the construction site, looks everything over, measures, considers, possibly meditates, and then decides: "Ah, ha, I shall build myself a cathedral with stucco walls, a great deal of glass, and stone decoration." He then extracts from his huge pile of materials the ones he has decided to use and turns his back on everything else. He will build the cathedral that his mind—his free will—his creative imagination—has instructed him to build.

The material he has left unused is for the moment extra, extraneous to his purpose, but it remains building material whether or not it's been used. Quite possibly, as he builds his cathedral, he will alter his plans somewhat and call upon materials that are not called for in his original blueprints. The *potential* for changing his structure is always there, for he started out covering every base he could think of, ordering enough material for every possible contingency. As he could not possibly know ahead of time what kind of cathedral had best be built, no one should fault him for being extravagant, for ordering more than he needed. He did *not* order more than he needed; he simply did not find use for all that he had. But maybe tomorrow, the week after next, or next year he *will* find use for it. Building the cathedral is his one task in life and will never really be finished until he dies.

Hair coloring is believed to be hereditary, one of the many traits ruled by our genes. Few babies, at least Caucasian babies, keep the hair coloring of the soft down covering their skulls at birth. The blond hair of early childhood tends to darken with age.

As a very young child I was towheaded, with fine straight hair that was almost white. Over the years my hair became golden blond and thickened to become wavy-curly. As I entered my teens it darkened to light brown, then to darker and darker brown. By the time I was forty it was so dark a brown as to be almost black and was steaked with gray. By the time I hit fifty it was completely gray. It has not yet lightened back to the near white it was to start with but I expect it soon will. That's quite a range of hair coloring for anyone to "inherit."

Eye coloring is another of the traits believed to be hereditary, determined by genes. It is reported that all Caucasian babies are born with blue eyes; many of these blue eyed babies will lose their blue eyes somewhere on the path to maturity. I was blue-eyed into my mid-teens, following which my blue eyes changed to green and I've been green eyed ever since.

The blue eyed blonde I was at twelve matured into a nineteen-year-old green eyed brunette.

How is this all accomplished under conventional genetic theory?

First of all, according to such theory, I inherited from each of my parents, one green eyed, the other hazel eyed, recessive genes for blue eyes. This happened by chance. I could as easily have inherited a gene (or genes) for green eyes from one, hazel eyes from the other. These recessive genes for blue eyes carried built in instructions (apparently) to change my blue eye coloring to green fifteen or sixteen years down the road. The DNA which contained the genetic instruction to give me blue eyes to start with, green eyes later on, was a short, "sensible segment" in a long strand containing mostly "gibberish."

The genes controlling hair coloring had to be given even more elaborate instructions: start white, change to bright blond, darken, darken, darken, then lighten, lighten, lighten, back (eventually, I feel sure) to the original white. These genetic instructions, too, came from *exons,* "sensible segments" of the DNA strand, with "nonsense" stretches on either side.

Is this really how it happened?

Let's consider another script:

SPECULATIVE GENETIC SCRIPT

Some mind or some force wanted to build a human body. Once the decision was made to cover the body with a white skin, it was decided to start with blue eyes. So along the strand of DNA containing instructions on how to color the eye—instructions for eyes of blue, blue-gray, green, hazel, brown—it zeroed in on the blueprints of its choice, ie, an eye colored blue. This was the particular blueprint wanted so the other blueprints—for green eyes, gray eyes, hazel eyes, brown eyes—are not called forth but are left for later consideration. The blue eyes satisfactorily express what the mind or force in the body wants to express for some fifteen years, but then, with a change in circumstances, green becomes more appropriate, so in recreating the molecules that form the eye, the word comes down to switch over, as quickly as can be comfortably done, to a different blueprint on the DNA strand, to the one that calls for green eye coloring and express that instead. Green proves to be satisfactory, expressive of what needs to be expressed, and no further changes (up to this point in time) are ordered.

With hair coloring the same procedure is followed. As the new hair grows in, a shift is made along the row of blueprints that determine hair color so that the hair coloring remains expressive of what the mind or force wants expressed. To begin with one small segment of DNA is expressed, then in time a shift is made and a new blueprint called forth. One shift after another occurs as the body ages and heads toward the grave.

****

In both of the above cases, it is taken for granted that DNA is the master builder, in charge of the blueprints. The first calls for only a small part of the DNA to be "sensible" while much of it is "nonsense." The second assumes that all of the DNA is "sensible," that every strand contains its alloted portion of blueprint whether or not every blueprint is called forth and constructed.

The crucial point of divergence is this:

In the former case, the DNA is viewed as its own master, in charge of itself.

In the latter case, while the DNA is seen as in charge of the building process, it is viewed as being strictly at the command of a force or will beyond itself. The genes are reduced to the role of the building blocks, lacking self determination.

SPECULATIVE COMMENT:

The genetic scientists have possibly been so enthralled watching DNA, the master builder, at work, marveling as the Boss synthesizes messenger RNA to match the chosen blueprints and sends this RNA on its way out to the cell, that they may have forgotten that behind most builders stands another, the one who visualized the structure to begin with and drew up the blueprints, one who may revise here and there to suit unexpected contingencies: the architect.

It is suggested that, in the building of the human body, free will, inspired by emotion, guided by the creative intelligence, is this neglected, so far undetected presence: the architect.

It is further suggested that it is the architect—not the DNA, with its mass of blueprints—who decides exactly which blueprints will be called forth and built, who has, in effect, the final say-so on all matters of construction.

This is just another way of restating the earlier hypothesis.

HYPOTHESIS:

Man has free will, this free will extends even into his genes, and he builds his own body to suit his own needs.

****

QUESTION:

What proof is there that man's free will has this much power, that such an Architect of the human body exists?

ANSWER:

There is no proof at all.

QUESTION:

If such an Architect exists, with the powerful electronic microscopes now in use, wouldn't this Force have been seen or detected?

ANSWER:

The human brain has been observed, tested, surgically operated upon, yet the mind has never been seen or detected. Is this presumptive proof that the human mind does not exist?

****

Genetics versus Envisionment.

Genetics: "the science of heredity."

Envison: "v.t. to contemplate in imagination, esp. as coming into being."

The suffix *-ment:* ". . . denoting an action or resulting state."

There apparently is no such word as *envisionment,* but it is here suggested that there ought to be, and that one of its definitions would be: "The process or action by which the human will brings into material being its physical form, the human body."

If there is such a process or action as *envisionment,* as defined above, how does it work?

SPECULATIVE COMMENT:

We all know that the human animal is extremely prone to imitation. Little girls don the high heeled shoes of their mothers, smear on makeup, and parade around as miniature "Mommies." Little boys slip their feet into Daddy's work boots, pick up his lunch pail, and stride manfully through the house. "Imitation is the sincerest flattery." *Nathaniel Cotton.*

But long before the young girl or boy is old enough to take on the clothes, manners and beliefs of her/his parents/loved ones, a human body has been forming—a body that began within the broad framework of a human mold but with little specification beyond that.

Most newborn babies look like other newborn babies. The parents of a newborn, if allowed to walk through a hospital nursery wherein were kept fifty or even twenty newborns, would be hard put to it to identify their own simply on the basis of appearance. Immediately following birth, newborns are nametagged to try to make sure the new mother takes home the child to whom she gave birth. Newborns are also footprinted as another hedge against inadvertent mix-up. The newborn Jones baby looks far too much like the newborn Smith baby and too little like his parents for much reliance to be put on appearance.

Despite this lack of resemblance at birth, the Jones boy quite possibly will grow up to be a carbon copy of his father, or, perhaps, he may grow up to look remarkably like his mother instead, while the Smith boy will quite possibly grow to strongly resemble one or both of his parents. How does this resemblance come about?

Standard genetic theory assumes that heredity causes it, that our genes program us to look like our progenitors. If we fail to follow this pattern and grow up looking more like the people next door than like our own family members, genetics readily supplies "logical" reasons for this: dominant versus recessive genes, plus or minus all the oddball hazards the genes are heir to. Any given organism can always be a throw-back to unknown forebears—still through the genes, of course—or maybe a couple of handy mutations occurred. Free will may come into play—*does* comes into play—when it comes to our posture, our stride, our facial expressions, our speech patterns, our religious beliefs, and our general behavior. It may determine whose clothes we dress up in when we're little and whose hopes and dreams we take on as we grow up, but it is assumed *not* to have any part in forming the underlying physical structure itself. In this area, in conventional genetic theory, the genes are still in control, *in interaction with the environment.*

But one of the major forces, if not *the* major force,in the immediate environment of the human gene is *the human mind with the human will.*

Genetics versus envisionment.

SPECULATIVE HYPOTHESIS AT WORK:

The human infant, as he opens himself up to his external environment, as he grows to depend upon, fear and love those who take care of him interacts with his inner environment, his mind/free choice/will, by taking on, in whatever measure suits him, the physical characteristics of those he fears and loves as he will later don their clothes, their speech patterns, their facial expressions, their religious beliefs, their prejudices and their dreams.

First the physical body itself, then all the rest.

This is a lifelong process, ending only when life in the physical body ends.

As we find new loves and painfully uproot the old, as we become joyful or disillusioned, as we unearth our buried fears and deal with them or leave them festering deep within, we modify our bodies in keeping with our needs. A daughter who may in no way resemble her mother physically at age eighteen may grow to be almost a carbon copy of Mama by the time she is fifty. A round, broad, adolescent face may change, as mine did, into a long, narrow, middle-aged face. Or a narrow face may grow to become a broad one. An eighteen-year-old daugher who, in her mother's wedding dress, could have posed for her mother's wedding pictures, may, by age fifty, have traveled far from the original. We reshape our features as we go, slowly molding them to express what's in our hearts; revise our facial features much as we broaden and change our vocabulary, alter our stance, our walk, our way. If we rebelled in childhood and at eighteen were a mutant—"How did *she* ever get into *this* family?"—by age fifty we are often very clearly back in the fold. If we lovingly followed our hearts in childhood so that on our wedding day we were Mama reborn, new loves may lead us far astray before we die. As our feelings emerge into the light and are expressed, as our fears are faced and die away, as our love contracts or expands, our wills reflect this and our bodies change. Life is not a static, gene controlled process, but a vital flow of feeling and will.

Envisonment.

Envision: to contemplate in the imagination; *-ment*: denoting an action. *Envisonment:* to contemplate in the imagination and bring into being.

If you don't like the body you have, look in the mirror and face the one who's responsible.

What is within us is also without.
What is without is also within.
Katha Upanishads

If we wonder what our true feelings are, feelings buried so deeply within that we may not even know they're there, we have only to look in a mirror, study our faces and our forms, to find a roadmap of our hearts.

SPECULATIVE SUMMARY:

If the human organism has the power to create itself in keeping with its own mind's eye view, how is this creation accomplished?

The same way most things are accomplished: through use of the raw materials at hand.

We know we inherit chromosome-bearing genes, that the DNA in the nucleus of our cells is the master builder, sending out orders as to which proteins are to be built. We know that DNA exists in what seems to be a greatly surplus amount. Whether the greater part of this DNA contains gibberish, as currently believed by scientists, or is instead sensible genetic potential so far unused as has been suggested herein, has not yet been proven one way or another.

We do know, however, that in the laboratory genes can be altered through chemical means, also through electrial means and through the application of heat.

Strong emotions in human beings bring about proven chemical changes in their bodies. Our brains work on electrical impulses. Bodily heat can vary depending on physical and emotional health.

Even if genes "rule" over us, we possess ample means at hand—chemical, electrical, temperature level—to alter them, to put our heart's imprint upon them, to bend them to our wills.

The cerebrum, the largest, most recently developed, most *human* part of the brain, has two hemispheres. The left rules over abstract reasoning and the verbal, the right over feeling and the spatial. The left is logical, the right creative. If, as an infant, we adore Daddy but don't yet have the words to express this, or if the feeling is too big for words ever to express, the right brain may be stimulated to express this love creatively, through imitation, the sincerest flattery: we take on the features and general bodily build of the one we adore.

On the other hand, if the feeling we naturally have for one of our parents is repressed, through fear, pushed out of the verbal left brain, pushed completely out of thought, it must go somewhere—love is too powerful a force simply to die away—so the unthought feeling splashes over into the creative right brain, where it will be expressed in bodily imitation. Murder may not always out, but love will, if not in thought and words than in face and form.

We can "take after" one parent more than the other so that an excess of love can find expression, or as the sole expression of a love we are too frightened to face.

Throughout our lives, as we work through our fears and allow buried feelings to surface, we modify our coloring, our facial features and our forms to express these changes. Nothing is left to chance: rather, the Free Will Architect has the final say-so on everything, from the instant of conception until the plunk of dirt on top of the coffin as the physical body gets lowered into its earthly grave.

What of adopted children who may grow up to resemble birth parents they have never seen rather than the adoptive parents with whom they have lived intimately all their lives?

What of them?

All adoptive parents and most adopted children know the truth of their relationship, that they are not blood kin.

If the only link you had with an unknown but secretly rather longed for past was a tarnished gold locket, one that you don't have the means to open but which may contain a greatly valued treasure if only you could open it, you would surely take extraordinary care of this locket hoping that someday someone would come along able to open it for you and reveal its contents.

Oftentimes the only link the adopted child has with his birth parents is his genes. In such a case, it makes sense that such a child would safeguard his sole inheritance as carefully as possible, would hold all gene tampering to an absolute minimum. Possibly the DNA, known possessor of great wisdom and skill, knows which of its vast assortment of genetic potential was actually expressed in the parents and in this way can follow the will of the child: "Take care, modify as little as possible!" In this case, the child's Free Will Architect wants nothing quite as much as to be ruled by the genes.

What of those people born with severe physical afflictions? Surely no one freely chooses this.

We all know that such severe afflictions *do* occur, so it must be matter of either choice or chance.

If chance deals blows of such severity to people who have not chosen, and do not deserve, the afflictions they have, then free wheeling evil is most certainly loose in the world.

This may be, of course; no one has ever proved that it isn't. But we would eliminate the primary evidence of evil in the world if we could prove that people are *not* chance victims of any physical afflictions they may have, even those present at birth.

Proof of this is not yet forthcoming, as far as I'm aware, but an argument in its favor is not difficult to come by.

How can we eliminate *chance* as the ruling factor when it comes to those people born with severe physical afflictions?

For those millions of people who believe in reincarnation, explanation can be given very simply in one word: Karma.

(*Karma:* "1. *Hinduism, Buddhism.* action, seen as bringing upon oneself inevitable results, good or bad, either in this life or in a reincarnation. . . .2. *Theosophy.* the cosmic principle of rewards and punishments for the acts performed in a previous incarnation.")

Throughout recorded history, far more people have accepted the theory of reincarnation than have rejected it. Those who believe in the doctrine find that it helps make sense out of what otherwise seems an unfair, unjust world. Injustice becomes only apparent, not real. Karma providentially explains all the negatives away. Those who suffer do so due to poorly lived past lives; those who live happily and successfully are reaping the reward for past lives well spent. It is neat and tidy, intellectually and emotionally satisfying, and, for those who accept and believe, goes a long way to "vindicate the ways of God to Man."

However, the fact that a majority of mankind has always embraced the doctrine of reincarnation, and that it makes beautiful sense to those who do embrace it, does not constitute proof of the doctrine's validity. Is there any way we can eliminate chance and elevate choice to a ruling position for those who view the notion of reincarnation as wrong headed wishful thinking and reject it?

In the absence of *karma*, is there any other explanation as to why some human organisms might choose to be born with, or deserve to be born with, severe physical afflictions?

It is a generally accepted feeling, among those who reject continued existence on earth in reincarnational form, that we can and do achieve a certain amount of earthly immortality through our children and our children's children. We were instrumental in creating them and often see ourselves reflected in them. Even if we ourselves are heading upward toward better things in heaven, or are hell-bent toward the eternal fires of damnation, some part of us still exists on earth in our children. For the atheist or agnostic for whom death is often seen as leading nowhere, for whom the grave is often seen as the end of it all, here at least, in physical form, in the love with which we conceived our children and nurtured them, is the one clear shot we have at immortality. As long as our children and their children survive on this earth, a tiny part of ourselves lingers on. We are not completely extinct.

That part of ourselves which we pass on physically to descendents is our *genes,* the long strands of DNA, the acknowledged Master Builder, the magical molecules with the blueprints.

Somewhere along the line can our genes decide that it is time to *choose* a severe physical affliction?

SPECULATIVE ANSWER:

"Sweet are the uses of adversity." Shakespeare.

Possibly we can't quite bring ourselves to see the uses of adversity as "sweet," but it is easy to argue that adversity does have its uses.

Pain is nature's warning system and those incapable of feeling pain are at a grave disadvantage, are, in fact, in a truly perilous state. Pain shouts at us: "Danger! Switch course!" If we can't hear the warning, if we are deaf to pain, we can suffer critical burns before we even know we've been singed. In shouting its warning to us, pain can save us.

When an imperfect organism is created, have the genes decided that it is time to choose a severe physical affliction, with its attendant pain, as a warning signal: *"Danger! Switch course!"?*

As the DNA goes to work generation after generation, building one new organism after another, what if a tilt sets in, what if the organisms thus created are expressing themselves in a way that is faulty? Sooner or later wouldn't correction occur, *karma not as applied to a reincarnating entity but action-reaction within the genes themselves?* If a given trait has been duplicated over and over again, one generation after another, only to put the organism in physical or moral danger, surely in time mutation would occur, a sudden change to bring the problem to an end.

This is not to imply in any way that physically afflicted newborns are the offspring of people who in one way or another lack moral character. The moral sense, and the genetic pool which produces the beings who possess it, can vary markedly from one human being to another. Some people don't flinch at assaulting, robbing, killing members of their own species, while others brood if they accidentally snuff out a gnat. We can feel guilty over destructive actions committed or because we are striving mightily toward a perfection not yet achieved, but *in either case the guilt is real.* How can we atone for our real and/or imagined transgressions? How does this powerful emotion express itself? If we can't face our own guilt and instead repress it, pushing it out of the verbal left brain into the creative right brain, what effect does this have on our genes?

> *Greater love hath no man than this, that a man lay down his life for his friends."*
> John 15:13

The newly conceived organism may, in love, choose to expiate the guilt splashing around in its genetic pool, may volunteer for the task of bringing a thorough cleansing to its genetic line. Sweet are the uses of adversity, and powerfuly sweet must be a love ready, willing and able to take on and atone for the real and imagined sins of its forebears.

In seeking immortality by having children, we may be passing on to them more than we think.

ODDITIES:

1. Man has a total of 46 chromosomes, paired sets of 23 each.

The gypsy moth, in contrast, has 62 chromosomes, paired sets of 31 each.

How does the gypsy moth rate *eight* more pairs than we have?

Surely it can't be that the gypsy moth needs greater genetic guidance than man does?

2. Monkeys have 48 chromosomes, paired sets of 24, one more pair than we have.

If genes truly make the man, maybe, if we could get our hands on one more pair of chromosomes and tried hard enough, we could make monkeys of ourselves.

6

THE DEMON THEORY OF DISEASE

Modern Medicine: The Same Old Black Magic That We Know So Well?

> If I had my own way I'd make health catching instead of disease.
>
> Robert Green Ingersoll

Throughout most of recorded history, magic and medicine have been inextricably intermixed.

For primitive man, disease is an affliction caused by the entrance into the body of something evil, a force possessing the power to cause mental, emotional or bodily ills. How else rid the body of these unwelcome intruders than through magical rites and prayers?

> When the even was come, they brought unto him many that were possessed with devils, and he cast out the spirits with *his* word, and healed all that were sick.
>
> *Matthew* 8:16

Today so-called civilized man shares with Jesus and his contemporaries and with primitive man this same conception of the cause of disease. The names differ but the underlying principle remains the same. The medicine man calls the external forces that invade the body *evil spirits* or *devils.* Modern, civilized man calls them *germs.*

Where primitive man used only his instincts and his intuitions to tell him what caused disease, we have advanced far beyond this. We can now put our devils under magnification and watch them do their nefarious work. Or can we? As we watch our devils at work, what are we really seeing? or not seeing?

Modern medicine classifies our wide range of physical disorders into three basic types based upon their etiologies, or causes:

1. Infectious diseases, caused by the invasion of evil spirits or demons, that is, by pathogenic (disease producing) organisms.
2. Deficiency diseases, caused by lack of something essential in the diet.
3. Neoplastic diseases, caused by some disturbance of the body's growth-promoting functions.

An even simpler classification breaks diseases down into only two groups: infectious (caused by demons) and non infectious (not caused by demons).

A catch-all term for our devil is *germ,* and we have quite as fancy an array of germs as primitive man has of evil spirits.

Germs: "1. a microorganism, esp. when disease producing."

Microorganism: "a microscopic plant or animal," ie, "so small as to be invisible or indistinct without the use of the microscope."

Germs may be classified into three groups:

1. Plants, such as bacteria and fungi.
2. Animals, such as protozoans and worms.
3. Virus.

Each of these general types consists of many subdivisions, there being numerous kinds of bacteria, fungi, protozoans, worms and viruses. Quite possibly we outdo primitive man—as well we should, with all our sophistication—in the number and variety of disease-producing devils we recognize and catalog.

These devils enter our bodies in ways that any child would be able to pinpoint. We can inhale the devil through our breath, ingest one with our food, or it can penetrate the skin or mucous membranes, most often through a break in one or the other. Once these devils have gained entry, they may or may not cause disease.

Sometimes they fail in their mission to harm us due to our bodily defenses against them. For those germs inhaled, we can sneeze or cough them out, or expel them with a runny nose. Against those vast numbers we take in with every bite of food, we wage constant chemical warfare. Saliva contains a powerful germ-killer that rids us of many of the enemy and those who get by this first line of defense are most often destroyed by hydrochloric acid in the stomach. The third mode of entry, through the skin, is not easily accomplished, though our skin can be breached even where there is no open cut or wound in such places as the hair follicles or sweat glands. Some viruses—the most powerful devils we are up against—can achieve the near miraculous: they can penetrate even the unbroken skin or mucous membrane. Against such power how can we possibly protect ourselves?

The fact that these devils often get past our defenses to set up residence wherever it suits them in our bodies still does not necessarily mean we will fall victim to disease. We can have numerous viruses flowing through our blood streams, for example, without ill effect.

Different germs cause illnesses through differing means. Some germs occupy cells and destroy them. Others produce toxins, or poisons, which do us in. Each germ is also considered disease-specific, which is to say that a given germ produces a given disease and no other. Tubercle bacilli never cause anything but tuberculosis, the polio virus causes polio and nothing else, et cetera.

Our modern demon theory of disease, as you can see from the above, is all very logical and neat, but *how valid—how scientific—is it?*

Science: "1. a branch of knowledge or study dealing with a body of facts or truths systematically arranged and showing the operation of general laws."

Our modern demon theory of disease is certainly "systematically arranged" but so is primitive man's demon theory. With all our high powered electronic microscopy, when it comes to bodily ills have we really advanced signficantly beyond primitive understanding?

In today's medicine are we really dealing with "facts or truths . . . showing the operation of general laws"?

If not—if it can be shown that we're not—then our present day germ theory of disease is based on the same old primitive supersititions and not on anything that can rightly be called science.

How scientific—how removed from magical beliefs and practices—is our current medical thinking?

Let's first take a look at that minor scourge of us all, the common cold.

All of us surely know that colds are contagious, that we "catch" them from one another. Your youngest child comes home from school feeling ill, and before you can say, "According to all the latest research, colds are caused by a virus, which is an infectious agent, any of a group of ultramicroscopic, infectious agents that reproduce only in living cells," everyone in the household is sneezing or coughing. Or you go to work feeling just fine only to spend the day sitting next to someone who has watery eyes and a runny nose. A couple of mornings later you wake with a slightly sore throat and an immediate grudge: damn Clarabella for doing this to you!

On the other hand, of course, it may not work that way at all. Your husband may come home from work feeling rotten, moody with the wretchedness of an incipient cold. All family members rally round—physcially as well as emotionally—hoping to cheer and comfort him. You feed him, kiss him goodnight, share his bed. For a week he suffers the indignities attendant upon a cold while all family members sympathize—but no one else in the family gets the infection.

Or you have a luncheon date with a favorite cousin who shows up coughing and runny nosed. Despite her cold, she enjoys seeing you and you spend several hours together laughing, talking, reminiscing. The following morning you wake up feeling great, and the morning after that and the morning after that. It is eleven months before you catch your next cold.

How contagious is the common cold?

The medical profession considers the common cold an infectious disease, yet all of us know that on occasion we can be exposed yet escape infection. How is this possible?

The common cold has been studied intensively by medical researchers. The infecting agent, believed to be a virus as it passed through filters, was named *rhinovirus,* "rhino" meaning nose. More than one hundred varieties of rhinovirus have been found; new strains "quickly appear and disappear. Moreover, other families of viruses can also cause colds, and there are even some colds for which no virus can be found."[1]

The accepted scenario for the full scale production of a cold runs something like this:

Someone with a cold sneezes or coughs, emitting into the air several million of his infectious germs. His hands, infected by contact with his eyes, nose or mouth, touch material objects around him, and upon these objects he leaves several million additional germs. A soon-to-be victim breathes in this virus laden air, allowing invasion number one. As her hands touch objects previously handled by the cold sufferer, the germs swarming over the objects get onto her hands. Sometime later her hands touch her eyes or nose, leading to a second wave of rhinovirus invasion. She has by now been thoroughly contaminated with the highly infectious agent which caused the sufferer's cold and which will soon cause hers.

The nose, invaded by the enemy germs, does its best to fight back; "the cilia lining the victim's nose and throat sweep away intruders at a rate of six hundred sweeps per minute,"[2] but this countermeasure is insufficient against such a "massive invasion of particles" which "entrenches itself in the warm moist terrain."[3]

From then on, the rhinovirus is in and the victim will soon be very nearly out.

Having established a beachhold in the nose, the viruses invade the cells, take over the cells' "manufacturing center,"[4] and begin reproducing. "As early as 24 hours after the viruses invaded the nose, the new generation of viruses make their escape"[5] from the cells they have taken over, used and "fatally wounded." The "viral progeny—90 million strong"[6] leave the dying cells behind and move to new, healthy cells.

The 90 million new viruses occupy healthy cells, take over to again reproduce themselves, and flood the victim's system with the second generation of progeny. The second generation follows its forebears, the enemy germs now number "close to a trillion, and for every virus the lymphocytes (of the body's defense-immune system) manage to kill, 10,000 new ones take its place."[7]

The poor victim now feels as though she were dying and almost wishes she were.

By the fifth day after the initial invasion, the rhinoviruses have clearly won, but now an odd thing happens. "The virus abruptly stops invading new cells,"[8] and the victim, instead of dying, rallies and recovers.

Why the virus, after fighting so cleverly and successfully, suddenly quits the battle and surrenders is not known.

This is, in simplified form, the generally accepted scenario. If it's valid—scientific—then it must of course be reproducible under controlled conditions. Only when the same result can be obtained repeatedly, without a single failure, from the same, controlled cause, can a theory be considered valid. With a sufficient number of successes, and with no failures, a theory eventually can be formulated as a "law."

If medicine has at last divorced itself from magic and become a science, its "body of facts and truths systematically arranged" must show "the operation of general law."

Is there a "general law" underlying the above scenario as to how one person catches that highly infectious disease, the common cold, from another?

Two researchers, Dr. Jack Gwaltney of the University of Virginia School of Medicine, and his collaborator, Dr. J. Owen Hendley, tried this experiment: "They placed volunteers in chairs around a small table, and those with colds coughed, sneezed, sang, and talked loudly, but only one of twelve susceptible volunteers caught a cold."[9]

One out of twelve—"susceptible volunteers."

Eleven out of twelve "susceptible volunteers" failed to become infected.

One out of twelve does not even begin to establish a trend much less indicate any general law at work. Scratch the virus-laden air theory of cold transmission. When eleven out of twelve volunteers, subjected to deliberately contaminated air, freely breathe in such air over a fairly extended time period yet escape all infection, someone's got to come up with a better theory than that as to how we infect each other with our colds, *if* we do.

In 1946 Christopher Howard Andrewes, an English doctor, "launched his all-out assault against the common cold."[10] He became the first director of the Salisbury Common Cold Research Unit, housed in the "Harvard Hospital and a large complex of buildings just outside of Salisbury,"[11] England. Volunteers were lured to the facility in order for the staff to study the detection and transmission of colds. By 1970, more than 17,000 volunteers had visited Salisbury.[12]

The volunteers were placed in isolation, two to a unit, and after several days, during which time physicians determined that none of them had colds, inoculations were given. "All the tests were double-blind: some of the volunteers received cold viruses, others a harmless saline solution, and neither the physician nor the subjects knew which until the results were tabulated."[13] The staff found "that they could reliably produce colds in thirty-five to forty percent of cases."[14]

Thirty-five to forty percent.

With direct inoculation of the agent (rhinovirus) believed to cause colds, colds could be produced in less than half of all cases. Sixty to sixty-five percent of the volunteers, after being injected with the agent believed to cause colds, *failed to come down with any infection.*

This is not only failure, this is failure on a pretty massive scale. Any theory that cold viruses, following successful entry into the body, cause colds should have been swept into the trash bin then and there. Viruses may be present when a cold occurs, they may be a contributing factor, but there is very clearly no direct cause and effect law at work *when the failure rate is over sixty percent.*

"Perversely, the Salisbury volunteers did not seem to catch colds from one another, either. When cold sufferers were paired with healthy volunteers, there was one cold in nineteen cases at first, then one cold in five, then none in four,"[15] a remarkably poor showing for what is believed to be, and in the family unit is often experienced as being, a highly infectious disease.

To sum up, the Salisbury researchers seem to have learned this: when it comes to colds, you can infect some of the people some of the time, all of the people none of the time, and a few of the people never—a fortunate few who seem to have a built in immunity.

Research had "established that when rhinoviruses were placed directly in the mouth or throat, infection did not readily occur, but that the eye was easily infected. But how were they entering the eye?"[16]

In 1977 Dr. Jack Gwaltney of the University of Viriginia School of Medicine and his collaborator, Dr. J. Owen Hendley, found that the eyes are infected through hand-eye contact, contaminated fingers touching or rubbing the eye. "A few colds are transmitted by airborne particles, but the hands clearly had it."[17]

Clearly had what?

The task or burden of transmitting colds from one sufferer to another, primarily through the eye.

This is an odd state of affairs. Viruses partial to the nasal passages are relatively powerless to cause infection unless their mode of entry has been through the eye. We can breathe in contaminated air with relative impunity and have a better than fifty percent chance of not catching a cold even when we allow ourselves to be injected with supposedly cold-producing viruses. If these viruses successfully breach our defenses and lay us low, they suddenly give up their winning fight a few days later and allow themselves to be washed from the body. Curious indeed.

After years of patient research, "most of what can be said about colds is what we do not know."[18]

Let me offer a speculative hypothesis.

HYPOTHESIS:

The rhinoviruses are powerless to cause any infection *on their own.* Often they enter the body and are swept out again with no ill effect. On those occasions that they invade healthy cells, begin feverishly reproducing and bring on the wide variety of symptoms umbrellaed under the term "cold," they do so BECAUSE THEY HAVE BEEN ORDERED TO DO SO. The moment the order comes through that they are to cease all such activity, they dutifully follow this order too.

Who is handing down the orders?

The same Architect (the human will, inspired by emotion, guided by the creative imagination) who oversees construction of the body stays on the job to supervise all necessary maintenance, including innovations and, when unavoidable, destruction.

We all know that repairs—renovations—cause discomfort and inconvenience. Nevertheless, they are meant to serve a purpose and invariably do, in the human body quite as much as in any other edifice.

Some force within the body—the force that we have called the Architect—wills the body's diseases, their onslaught, their severity, their duration, and whether or not recovery will occur.

****

Let's take a look next at the tubercle bacilli.

Tubercule bacillus: "the bacterium, *Mycobacterium tuberculosis,* causing tuberculosis."

There it is, in plain English, straight from the dictionary: the tubercle bacilli are the bacteria that cause tuberculosis.

However, repeated tests have shown that, by the age of fifteen years, nearly everyone living under urban conditions has become infected with tubercle bacilli.

We have all been infected, almost without exception, yet only a relatively small minority of us has the disease. Therefore, a clear distinction must be made between being infected and being diseased.

The fact that this distinction *must* be made is positive proof—all the proof that any reasonable, open minded person needs—that tubercle bacilli do *not* cause tuberculosis.

Cause: "a person or thing that acts, happens, or exists in such a way that some specific thing happens as a result; the producer of an effect."

If tubercle bacilli "cause" tuberculosis, if they are "the producer of an effect," then that effect must follow not in a small minority of cases but without fail. If the "specific thing" does not happen as a result, then we have no "cause."

Were these bacteria truly the "cause" of tuberculosis, every body infected by the bacilli would have the disease.

No distinction is possible between being infected and being diseased if these bacteria "cause" the disease of tuberculosis.

SPECULATIVE COMMENT:

Rhinoviruses do not have the power to cause colds.

Tubercle bacilli do not have the power to cause tuberculosis.

To extrapolate from these specifics to a generalization:

The modern day demon theory of disease—the germ theory of disease—does not adequately explain the etiology of infectious diseases.

To correct this:

HYPOTHESIS:

While germs play a part in the production of disease, *no germ has any power to cause disease without the active acquiescence of the "victim."*

When it comes to his diseases, man holds the reins of power, not the invading germ.

To phrase it another way:

The human body is constructed in such a way that it can survive only when it receives frequent nourishment.

For the body to be nourished, two conditions must be met: there must be an intake of food, and the body must utilize the food taken in.

A superabundance of food will not keep a body alive and healthy if the mechanisms within the body to process and utilize the food have broken down.

In like fashion, when it comes to infectious diseases, two conditions must be met: there must be an invasion of germs (this probably holds true even for those colds for which no virus has yet been found, or possibly not; time should tell), and the body must put these germs to work to create disease.

Disease, like bodily nourishment, is cooperatively produced. If there's no cooperation, there's no disease.

****

With this hypothesis in mind, let's consider a once dreaded affliction, smallpox.

It was well known in the English countryside, in the 18th Century, that milkmaids who became infected with cowpox from the infected cows they milked would thenceforth be immune to smallpox. In 1774 Benjamin Jesty, an English farmer, took cowpox matter from a pustule on the nipple of a cow and successfully inoculated his wife and two of his children.

Dr. Edward Jenner (1749-1823), aware of the affinity between cowpox and smallpox, experimented first on his own son, Edward, Jr., born in 1789. He "injected his baby son with the virus of pox found on a pig. . . On the eighth day the baby became sick . . . several small pustules appeared. It is not known how long Jenner waited, but later he inoculated smallpox matter in little Edward's arms five or six times—without producing the slightest inflammation. . ."[19]

On May 14, 1796, Jenner scratched matter from cowpox sores into the arms of a healthy boy, James Phipps, who a week later came down with cowpox. Six weeks after the boy's recovery, Jenner "inoculated the boy with smallpox from a pustule on the body of a patient... Nothing happened... Jenner tried it again several months later. There was never another pustule, much less any general eruption of the body."[20]

Dr. Jenner called the cowpox material "vaccine" and the method for its use "vaccination," both words being derived from the Latin *vaccinus,* meaning cow.

The use of Dr. Jenner's vaccination spread throughout Europe, then to America. In those areas where vaccination was known and used, smallpox was never again a serious epidemic disease.

This is truly a magnificent success story. Smallpox was at one time a scourge; now, in those nations which mandate compulsory vaccination, it is a rarity. For this, we all owe a debt to Dr. Edward Jenner, also the milkmaid who drew his attention to the immunity from smallpox conferred upon those who had had cowpox, and even more to the young boy, James Phipps, picked apparently at random by Dr. Jenner to be experimented upon. ("In Jenner's words: 'I selected a healthy boy, about eight years old, for the purpose of inoculation for the Cow Pox.'")[21] Had Dr. Jenner made some unfortunate error in calculations, the boy James, into whose body a "virulent infection" of smallpox was later introduced, might not have lived to be inoculated a second time. But the boy *did* live; even more, he escaped smallpox entirely. On the basis of this one success, Dr. Jenner published his findings and in short order became both famous and wealthy. Vaccination was on its way and the defeat of smallpox imminent.

Even prior to the above happenings, inoculation against the dread disease smallpox was not unknown in England. In 1721, Lady Mary Wortley Montagu returned to England from Turkey, where her husband Edward had been ambassador, and attempted to introduce into her home country the "heathen" practice she had learned from the Turks.

Every September while in Turkey Lady Mary had "watched a group of professional old women do the rounds of houses in Constantinople. Their purpose was to practice a well-tried piece of preventive medicine."[22] Each of the women "would carry in a nutshell a small sample of pus she had collected from a victim of a mild attack of smallpox."[23]

A vein would be scratched open on the limb of a customer, the practioner would then "dip her needle into the pus in the nutshell, smear it on the open vein, cover the wound with a piece of nutshell and bind it."[24] The process was known as "ingrafting."

The children subjected to this inoculation (presumably the adults too) would remain healthy for eight days, then would suffer from a fever which might keep them in bed for two or three days. But "in eight days' time they are as well as before their illness. . . Every year thousands undergo this operation, and the French ambassador says pleasantly, that they take the smallpox here by way of diversion, as they take the waters in other countries. There is no example of anyone that has died in it; and you may believe I am very well satisfied of the safety of the experiment, since I intend to try it on my dear little son."[25]

In 1718 Lady Mary lived up to her word, and "Edward Wortley Montagu was the first recorded Englishman to experience this form of inoculation against smallpox."[26] He survived the experiment and was thereafter protected against the dread disease.

Both the Turkish practice of "ingrafting" and Dr. Jenner's cowpox inoculation were based on the observed fact that people who have had certain diseases rarely get them a second time.

Smallpox was one of these once-in-a-lifetime diseases. Those who survived the disease were, in most cases, immune thereafter. Modern medicine ascribes this immunity to the antibodies manufactured by the body in fighting off the initial attack. After the disease has run its course, the defense erected against it remains strong enough to ward off completely any subsequent assault. This is not an invariable rule, however, there being numerous exceptions to it. Dr. Jenner's successful vaccination showed that cowpox and smallpox were closely enough allied that the antibodies which remained following the cowpox attack could ward off smallpox.

Observation of the human condition has shown that there is not only the acquired immunity mentioned above, but that in addition a fortunate few have a natural immunity to certain diseases. When smallpox was devastating the civilized world with recurring epidemics, on occasion as many as seventy-five percent of the population would be stricken, but even this horrendous percentage leaves *twenty-five percent untouched.* Possibly a small percentage of this group suffered no exposure. Another percentage had surely acquired immunity from a prior onslaught. But there remain a fortunate few who, though exposed and not having any acquired immunity, nevertheless escaped the disease. Such fortunate people possess a "natural" immunity.

Where immunity is not total, there exists a variation in the amount of immunity. To put it another way, there are degrees to which different individuals are susceptible. Experience has shown that in any smallpox outbreak, the most severe cases occur first, those with the least immunity, the greatest susceptibility, being stricken first, followed by those with a somewhat greater immunity whose experience of the disease will be less severe.

Along with varying degrees of immunity which can cause a lesser or greater severity of illness, the disease itself can appear in a milder or more severe form. It will be noted that in the Turkish inoculation program, the women medics who carried on the ingrafting program secured pus from "a victim of a mild attack." In 1789 Jenner "described an outbreak which occurred about seven years before in Gloucestershire, 'of so mild a nature that a fatal instance was scarcely ever heard of. . . I watched its progress upward of a year without perceiving any variation in its general appearance. I consider it then as a variety of smallpox.' "[27]

This milder version of smallpox has occurred in other parts of the world also, and while some medical authorities consider it a closely allied but different disease, most believe it a mild type of smallpox.

The immunity to smallpox conferred by vaccination has been found to diminish with time, the amount of loss varying with the individual. In order to boost protection back to the highest level, re-vaccination has become accepted practice.

Another variable: Back when epidemics raged unabated, experience showed that the pestilence hit hardest where living conditions were crowded and unsanitary. Common sense would suggest that possibly two factors were at work here: one, the more crowded the living conditions, the greater the exposure; and two, people living in slums may be less well nourished and generally somewhat less healthy, which could account for a greater susceptibility. Whether these two factors actually accounted for the difference, or whether other factors were also involved, is not known.

To sum up, we can make these statements about smallpox:

A virus causes smallpox—but only in the absence of natural or acquired immunity.

Smallpox is a dreadful, often fatal affliction—except when it appears in much milder form.

Those who have once had smallpox are immune thereafter—except in a few rare cases.

Vaccination affords protection—but the level of protection decreases with time, the amount of lost protection varying with the individual.

Uncrowded, sanitary living conditions afford some protection—but far from total immunity.

****

In these statements about smallpox, we come up with quite a number of exceptions and variables. Even the basic statement, "A virus causes smallpox," has to be qualified, for the virus does *not* cause smallpox in those who have a natural or acquired immunity.

To put the above information into proper prospective, let's apply the same exceptions and variables to a known physical law, one that we all have great familiarity with as we live with it every day of our lives: Newton's law that all bodies in the universe attract one another, commonly called the Law of Gravitation.

Ten men are lined up along a high roof and are ordered to jump off at a given signal. When the signal is given, all ten jump. Nine fall to the ground, but the tenth one hovers indefinitely in the air. For some reason he has a "natural immunity" to the Law of Gravitation.

Of the nine men who jumped and crashed to earth, those who jumped the most quickly received the greatest injuries. The first three to hit the ground were fatally injured, the next three were severely but not fatally injured, the final three were the least severely injured. The man who was the slowest to jump escaped with the least amount of injury.

Ten other men jump off the same roof. This jump proves to be much "milder." No one dies; no one is severely injured; apart from a couple of twisted ankles and some bruises, everyone escapes unscathed.

Ten of those who have jumped previously jump a second time. This time only one of them falls to the ground. The other nine are now immune to the Law of Gravitation as they have already experienced it. The tenth, the one who fell, unfortunately is the rare instance of someone who failed to gain immunity after an initial experience.

Ten men who years before jumped off a low roof now jump off a high one. Some fall very rapidly, others more slowly, some hover and don't fall at all. The immunity to the Law of Gravitation which they gained from their low jump has decreased at a variable rate, some of them losing protection faster than others. Therefore, despite the Law of Falling Bodies promulgated by Galileo, they fall toward the ground, if they fall at all, at different rates.

Ten men who have been watching the jumps split into two groups and head home. Five are from a crowded, unsanitary section of town, the other five from an uncrowded, far more sanitary section. Each group of five jumps off a roof in its home area. The five who jump in the poor part of town fall more quickly, and are more severely injured, than the five from the more affluent area, as the Law of Gravitation operates with greater swiftness and severity under slum conditions.

The above is meant to make clear the startling difference between an established law of physics and those beliefs which pass for "law" in the field of medicine.

The United States could not possibly have put a man on the moon if our scientists had been dealing with laws of motion as uncertain, subject to as many exceptions and variables, as this "law" about the causation of smallpox: "Smallpox, a serious often fatal disease, is caused by a virus."

In the physical sciences, if there is one proven exception to a law—if a given law fails even once—such a "law" is immediately written out of the books and a fresh start taken.

With an uncounted number of exceptions to the "laws" that certain germs "cause" certain diseases, the "laws" are still upheld, not only upheld but almost universally honored, as modern medicine holds to its "laws" as religiously as witchdoctors hold to theirs.

SPECULATIVE COMMENT:

The "law" that, "Germs cause disease," or, turned around, "Diseases are caused by germs," or, narrowed down, "Infectious diseases are caused by disease-specific germs," should be gratefully interred, with only the good the idea did us living after it.

Germs do *not* "cause" disease in any unvarying, this-is-the-law fashion. When a given effect often fails to materialize, there is no direct causation. We must look more deeply into the matter before we can declare a direct, unvarying cause of disease.

To restate an earlier hypothesis:

HYPOTHESIS:

Germs are *not* the cause of disease, though they may, in some diseases, function as helpers.

Disease, like bodily nourishment, is cooperatively produced. If there's no cooperation—no verifying order from the Architect, the human will—there's no disease. A force within the body wills the body's diseases, their onslaught, severity, duration, and whether or not recovery will occur.

****

Before we plunge into any discussion of this Architectural Force and attempt to explain how it operates and why it wills disease upon the body it inhabits, let's take a look at another virally "induced," infectious disease, yellow fever.

Yellow fever "first traveled west on slave ships in the seventeenth century."[28] It was "the most dreaded of all plagues."[29] Victims turned lemon yellow and "the terrifying black vomit started."[30] In 1793 the fever swept into Philadelphia. "In two months the epidemic killed one of every ten Philadelphians."[31] Then, after several months, "the plague departed, as abruptly as it had come. People observed with relief that the frost had seemed to kill it."[32]

During the 19th Century, despite all the precautions taken, the epidemics worsened, to where "it became a terrible menace in ports from Boston to New Orleans."[33] It was rife throughout the Carribean, in Haiti routing Napoleon's troops and, in Panama, along with malaria forcing the French to abandon construction of their canal.[34]

During the Spanish-American War, yellow fever raged among the American troops in Cuba. When the war ended, "Major Walter Reed was dispatched to Cuba to study yellow fever. The commission arrived in Cuba in the summer of 1900."[35] For some years a physician in Havana, Dr. Carlos J. Finlay, had been claiming that mosquitoes transported yellow fever, but he had not been able to convince anyone of this.[36]

When the four man American commission headed by Reed, after intensive study of eighteen yellow fever victims, "failed absolutely to find any cause"[37] for the fever, they went to call on Dr. Finlay to listen to his "fool theory" about mosquitoes.

Dr. Finlay turned over to the Commission "some little black eggs shaped like cigars and said: 'Those are the eggs of the criminal!' "[38] Jesse Lazear, one of the Commission members, "put the eggs into a warm place to hatch into wigglers, which presently wiggled themselves into extremely pretty mosquitoes, with silver markings on their backs—markings that looked like a lyre."[39]

To prove that mosquitoes carried yellow fever, experimental animals were necessary, but at that time it was erroneously believed that no animal was susceptible to the disease except for man. For the mosquito theory of transmission to be tested, man would have to serve as the experimental animal. "But give human beings yellow fever! In some epidemics—there were records of them!—eighty-five men out of a hundred died of it, in some fifty out of every hundred—almost never less than twenty out of every hundred. It would be murder!"[40] Nevertheless the decision was made to go ahead, using volunteers.

Lazear took his pretty, "silver-striped she-mosquitoes" and "went down between the rows of beds on which lay men" stricken with yellow fever, allowing the mosquitoes to bite and suck their fill.[41]

These she-mosquitoes, once they'd "digested their meal of yellow fever blood," were applied to the limbs of eight volunteers and allowed again to bite and suck their fill. But, alas for the theory of mosquito transmission, all eight stayed "fit as fiddles," with nary a symptom of yellow fever among them.[42]

Dr. James Carroll, another member of the Commission, mindful of Major Reed's orders, prior to his departure for Washington to make a report on their work, to "Try mosquitoes!" insisted that they try again. "He told Lazear to bring out the most dangerous mosquito in his collection,"[43] and had himself bitten by it. Four days later he was stricken by the fever, "the first case to come down with yellow fever after the experimental bite of a mosquito."[44]

When Dr. Carroll showed signs of the fever, the same mosquito which had bitten him was set up to bite another volunteer, an American private soldier, William Dean, and he too was stricken.

Ten men experimentally bitten, and two of them stricken.

It was later determined that the first eight men had failed to succumb to the fever not due to any natural immunity but because the mosquitoes used to bite them had bitten the yellow fever victims too late in the course of their disease.

Although the mosquitoes were gorged with yellow fever blood, it was not infectious blood.

When Dr. Carroll and the soldier William Dean were successfully given the fever, "the insect causing these two infections had been hatched in the laboratory and had *fed twelve days before upon a yellow fever patient,* who was then *in the second day of the disease.* This, then, was the proper combination of these two factors necessary for a successful inoculation."[45]

Once Dr. Carroll and William Dean had been successfully inoculated, further controlled tests were necessary to prove that the mosquito bite, and it alone, had caused their disease.

New volunteers were isolated, kept under isolation long enough to make sure they weren't already infected, then bitten by "dangerous" mosquitoes. Other volunteers were put into small houses, fourteen feet by twenty feet, into which were brought "pillows, soiled with the black vomit of men dead of yellow fever," sheets and blankets, "dirty with the discharges of dying men past helping themselves." Every possible source of contagion, other than the mosquito, was tossed in, with later volunteers sleeping "in the very pajamas in which yellow fever victims had died."[46]

When the mosquito was allowed in to bite, yellow fever resulted, even under the cleanest, most sanitary conditions.

As long as the mosquito was barred, it did not matter what other elements were present—how soiled and foul smelling everything was—no yellow fever resulted.

But what if the men sleeping on the pillows and sheets, and under the blankets, and in the pajamas of men who had died weren't *susceptible?* "Maybe they were naturally immune!"[47] So two of the volunteers who had inhabited the stench-filled little house were subjected to a further test: one had "virulent yellow fever blood" shot under the skin, the other was bitten by mosquitoes "who had fed on fatal cases of yellow fever."[48] Following this, both were stricken, so no natural immunity had protected them earlier. In these kinds of controlled tests, Dr. Carlos Finlay was proved right: His "little black eggs" *were* "the eggs of the criminal." Mosquitoes carried yellow fever from one victim to the next.

Once this had been established, there was an all out campaign to rid Cuba of the *Aedes aegypti,* the highly domesticated carrier mosquito. Less than a year after the intensive cleanup campaign was undertaken, "Havana was free of yellow fever for the first time in 150 years. After similar campaigns, yellow fever disappeared from the United States.... By 1925, yellow fever also disappeared from Mexico and Central America, and a cleanup campaign was launched in Brazil."[49]

This is a beautiful, soul-stirring success story if ever there was one, so much so that Paul de Kruif could write triumphantly, in 1926, "there is hardly enough of the poison of yellow fever left in the world to put on the points of six pins...."[50]

While de Kruif's jubilation was natural enough, his statement was, unfortunately, a bit premature. By 1928, "yellow fever was once again an urgent concern."[51] However, this resurgence of the old fever is another story and not one that will be explored here.

In later experiments it was determined that yellow fever, like smallpox, is "caused" by a virus.

There are other similarities to smallpox. Once is usually enough, one attack assuring most survivors of lifetime immunity. The degree of susceptibility varied also. During the course of the controlled studies in Havana, two volunteers, a soldier named Kissenger and a civilian clerk named John J. Moran, were put into preparatory quarantine and later bitten by infected mosquitoes. Kissenger was stricken by yellow fever, Moran was not. Yet that Moran had only a partial "natural immunity" was proven later when he underwent additional, repeated bitings and this time did succumb. Of five "mercenary fellows,"[52] Spanish immigrants who agreed, for the sum of $200 each, to "volunteer" for the experiment, four were stricken after being bitten while one was not.

The recovery rate for the Commission's volunteers was phenomenal. The first two to be successfully inoculated, Dr. Carroll and William Dean, both recovered. The next volunteer, Kissenger, recovered; Reed and Jernegan recovered. John J. Moran, once the mosquitoes finally got him, recovered—*this with a disease which often claimed fifty to eighty-five percent of its victims.* The only early casualty was Jesse Lazear. Although Lazear was a member of the Commission and one of the first to volunteer to be bitten, his death did not result from a planned inoculation. After the initial failure in which he participated, he became skeptical of the theory that mosquitoes transmitted yellow jack and in consequence didn't bother to brush away a mosquito which had settled on the back of his hand as he visited a ward filled with yellow fever victims. A few days later he came down with the fever and six days later died.

As with smallpox, yellow fever could produce a mild illness, a severe one, or one that was fatal.

"'In my own case,' said Carroll, 'produced by the bite of a single mosquito, a fatal result was looked for during several days. I became so firmly convinced that the severity of the attack depended upon the susceptibility of the individual rather than on the number of bites he had got, that on October 9, 1901, at Habana, *I purposely applied to a non-immune eight mosquitoes (all I had) that had been contaminated eighteen days before. The attack that followed was a mild one. . . .*'"[53]

This is the testimony of a man on the scene, directly involved, "the first case to come down with yellow fever after the experimental bite of a mosquito."

Natural immunity—acquired immunity—varying degrees of susceptibility, just as in smallpox or other infectious diseases.

Some years later, when yellow fever had reappeared and had become once again a subject of intensive research, it was decided to try to culture in the laboratory "the vicious Asibi strain," which was the "most lethal yellow fever virus known."[54]

And where had this "vicious" strain come from?

From a twenty-eight-year-old black man named Asibi, who "had been seen in a village about a hundred miles north of Accra on the Gold Coast. When Rockefeller Foundation doctors first encountered him, Asibi was resting with his head in his hands, feeling very ill. They diagnosed yellow fever and drew some blood.'[55] The virus that was later isolated from this blood was not only the "most lethal yellow fever virus known," it was also "pantropic: deadly to both viscera and nerve tissue."[56]

So what of this poor man Asibi who had harbored a virus of such lethal intensity? "Returning to the village several days later, the doctors learned that their patient had recovered and gone back to work."[57]

Would that we all had the disease fighting potential of Asibi!

Why didn't the doctors stick around to study the man, to find out from whence cometh his strength, instead of expending themsevles on a study of the "most lethal yellow fever virus known" that caused its possessor only a few days of missed work?

QUESTION:

Does each of us have the potential to become as disease-resistant as Asibi?

ANSWER:

Yes, if we take the trouble to get to know, and to get on good terms with, the Architectural Force within us which oversees our physical history and okays or vetoes disease.

7

HEALTH, GOOD AND OTHERWISE

If We Prefer the Good, Why Do We So Often Succumb To the Bad?

> Life is not a spectacle or a feast: its a predicament.
> George Santayana

When we come to consciousness and find ourselves in a human body, how can we best deal with the predicament this poses?

What is the best way to assure ourselves of comfort as we work within our physical limitations? the best way to avoid the long list of ills the flesh, blood and bone are heir to?

In brief, how can we maximize our pleasures and minimize our pains?

We know that some people are fortunate enough to have a natural immunity to infectious diseases. Others have little susceptibility to them; they have to be repeatedly exposed to infection before the infection takes. When it comes to non infectious afflictions, ranging all the way from a minor headache to terminal cancer, the degree of resistance also varies widely. Some people can remark in all truth that they haven't been sick a day in their lives. Others spend half their lives in bed or dragging themselves painfully out of bed. A blessed few live what seem to be entirely pain-free lives, are rarely troubled by even a minor ache. Others keep enough medication on hand to stock a pharmacy, pop pills on the slightest provocation, yet remain pain-afflicted even when heavily drugged.

How can we align ourselves with the pain-free—if that is our wish—and avoid the anxiety and discomfort of disease?

Disease: "1. *Pathol.* a condition of the body in which there is incorrect function resulting from the effect of heredity, infection, diet, or environment; illness, sickness, ailment."

Is there anything approaching a sure fire way to avoid disease, to avoid being ill, sick or ailing?

For starters, we can of course pay attention to what the doctors and dieticians tells us: eat properly, don't overdo, get plenty of rest, learn how to relax and avoid undue stress. Don't smoke. Drink alcohol only in moderation. Exercise regularly. Maintain a proper body weight. Have regular medical check-ups. With this regimen, if you're lucky, you'll feel reasonably well and will stay fit.

But what if you aren't lucky?

What if, in spite of your best efforts to follow all of the above injunctions, your body seems to have it in for you and malfunctions frequently out of what seems to be nothing but plain onriness or spite?

We can all look around and see those who break many or all of the above mentioned health rules yet somehow, for some unfathomable reason, get away with it, while others adhere religiously to them and rarely experience the joyful vigor of bursting good health. Why? And how can we align ourselves with the former and break rank with the latter? Is there some important health rule not generally known that, if only we followed it, could assure us our best shot at staying happily fit?

Consider this situation. You are due to run in a marathon race, twenty-six grueling miles. For years you have trained as a runner, you love running, you *want* to run in this race. But at the last minute a problem arises.

You put on newly purchased running togs and nothing fits. The sweat band you wear around you head is so tight it will guarantee a headache. Your shirt is so tight you cannot move freely, your shorts even tighter. Worst of all, your running shoes are at least three sizes too small and will all but cripple you. In your overly tight clothes it hurts you to breathe, your limbs are constricted, and your feet—oh, sweet mother, there is no way you can run in these shoes! But the race is about to begin, the gun goes off, you have no choice. Off you go, in your ill fitting clothes, struggling forward as best you can.

Or let's say the problem is of exact opposite dimensions. You dress for your marathon race only to find that your newly purchased clothes are several sizes too large. The sweat band meant to stay snugly in place on your brow is so loose it refuses to stay where it belongs, instead keeps dropping down over your eyes. Your shirt is so many sizes too large that it will flap wildly in the breeze, the shoulder seams being down at your elbows; your pants are so loose they pose an ever present danger of falling down. But worst of all are the shoes, gunboats so long and wide that even your tightly curled toes may not succeed in keeping them on. On, dear mother, there is no way you can run in these clothes, in these shoes! But the race is about to begin, the gun goes off, you have no choice. So off you go in your ill fitting clothes, struggling forward as best you can.

Fanciful? Yes. No marathoner would ever allow herself or himself to be caught in such an outlandish predicament. But when it comes to going forward in life, *this is the predicament most of us are in most of the time. We dress ourselves in clothing that does not fit.*

To return to our imagined runner.

In the too tight clothes, our runner is going to experience pain: a headache from the constricting sweat band, shortness of breath from the too tight shirt, abdominal discomfort from the too tight shorts, and her/his feet will soon hurt dreadfully from the crippling effect of too tight shoes.

In the too loose clothes, our runner will fare no better. If the sweat band drops down to blind the eyes, this could precipitate a fall. The too loose sleeves of the shirt can catch on nearby objects. Falling shorts can not only cause embarrassment but are quite likely to trip one up. And the oversized shoes offer far more threat to safety than protection. All too easily the runner could find herself or himself unintentionally pitching forward into a sprall on the road leaving behind unmoving shoes. A sorry predicament indeed.

But let's leave behind all thought of our poor benighted runner and consider what is happening to the ill fitting clothing.

The too tight clothing, subjected to the runner's attempts to loosen the constriction it imposes, is going to be under constant strain. If the runner rises above her/his pain and keeps running, the overly tight clothing will gradually loosen—until it bursts a seam or two. In time it will tear or split to where it becomes unusable as clothing unless/until repaired. Damage has put an end to its proper functioning.

The too loose clothing will also be subjected to immediate, excessive wear and tear. Flapping sleeves catch all too easily on protruding surfaces, which may tear them. If droopy pants fall, they can be dragged along a rough surface and shredded. Shoes too large to keep on, stepped out of and abandoned, have no way to protect themselves against being squashed down by others or run over. Soon all of this clothing too will become unusable without repair. Damage has put an end to its proper functioning.

SPECULATIVE COMMENT:

Our primary predicament as human beings is that we outfit ourselves improperly. All too often nothing we own fits properly and we don't know how to make adjustments to bring about a proper fit.

****

The "normal" or average human being has a body, a mind, has feelings, dreams, et cetera.

Who is it that has all of these things?

I do, you do, she does, he does, they do.

Each of us has a deep down sense of selfhood—*"I am I."*

This "I" has a body but the "I" does not equate with the body. We can, if we're unfortunate, lose one or more limbs, but the sense of self remains. As long as the body lives and there is consciousness, there is a sense of "I." Even when the "I" becomes confused and thinks of itself erroneously as Cleopatra or Napoleon, or as a cat or panther or horse or chair or table or hole in the ground, there is still an "I." This "I," this self, this whatever-it-is, has a seemingly inexhaustible supply of thoughts and feelings as it inhabits, or possibly only seems to inhabit, the human body.

In earlier chapters I have referred to this indwelling force—the self that possesses without being possessed—as the human will, or the Architect. It is not the body as such nor the conscious mind as such, but is the force or self or entity or spirit or soul or whatever-one-wishes-to-call-it that *has* a conscious mind, with its attendant thoughts and feelings, and also *has* a physical body. It is the bedrock "I am I" within each of us.

For the purposes of the discussion to follow I will refer to this force, possessor of the mind and body, as the *spirit,* an arbitrary choice over such possible terms as the self, the soul, the will, the ego, et cetera. The deep indwelling voice in each of us that says, "I am I," is that to which I am referring, and for which I will use the word *spirit.*

SPECULATIVE COMMENT:

A proper marathon can be run when the clothing worn fits the runner. The clothes worn and the body which wears them both belong to the physical world.

The spirit in each of us possesses a mind, and no matter how tied to the physical body both spirit and mind may be, neither is expressed in the physical world in quite the way the clothes and the physical body are. Scientists cannot put either the mind or the spirit under a microscope and examine it. In this sense, both spirit and mind are relatively non physical.

GOOD HEALTH HYPOTHESIS ONE:

Just as, on the physical plane, for ease of movement and healthy, happy functioning, the clothes must fit the body, on the non physical plane the mind must fit the spirit.

Good health exists when the mind is a proper fit for the spirit; the better the fit, the more radiantly joyful the health.

In life, most of us run a much longer race than a twenty-six mile marathon. During the course of the race—as in any long race—the power in charge, the spirit, is forced to make increasingly stringent demands. The mind must keep up with these demands or strain results. With prolonged strain, damage occurs, just as it does to the ill fitting clothes of our imagined runner.

GOOD HEALTH HYPOTHESIS TWO:

The mind stops being a proper fit for the spirit when it becomes polluted with too many unexpressed negative feelings.

Just as physical waste material must be expelled from the body or we'll become ill, possibly even die of our own accumulated poisons, so must negative emotion be dealt with and discharged. If there is a build up of repressed fear, anger, guilt or resentment to where the spirit, offended by all this poison, feels pained, the spirit will demand a cleansing. If the mind refuses to handle this cleansing, the body is forced to take on the job—discharging the negatives through a process we call disease.

GOOD HEALTH HYPOTHESIS THREE:

Bodily strain, with possible resultant damage, occurs precisely where the indwelling spirit experiences lack of a proper fit.

If our shoes are too tight, our feet hurt. Our hurting feet let us know exactly where the tightness is, where the strain is.

In identical fashion, any physical discomfort we experience pinpoints the area in which the spirit feels pain.

****

To see how this works, let's have a look at the common cold.

Different people experience colds in different ways. Some begin with a severe sore throat, then once the throat burning dies down, all other symptoms are mild or non existent. Others may experience a mild sore throat, or no soreness at all in the throat, but become afflicted with what seems an endless nasal flow, a veritible upside down fountain of ill-smelling splash. Others may skip a heavily runny nose but suffer painful head and chest congestion. Some may get through the cold itself with only minor discomfort, then experience a hacking cough that lingers on for what seems like forever.

These are only some of the various ways we can experience what is known as the common cold.

There is not only a wide variance in the way cold sufferers suffer colds, the same person may experience colds in different ways at different times. Today a burning sore throat, six months ago an endlessly running nose, a year ago a bad night cough—each cold may highlight itself in a different fashion. Medical researchers tell us that over a hundred different "cold-causing" viruses have already been isolated, which may possibly account for some of this variance—for those who accept the germ theory of disease—but any such germ theory leaves a great deal unexplained.

I may experience a cold largely with a wretchedly burning throat, then hand the cold onto my husband, and the same cold germ—presumably the same cold germ—"causes" in him a very mild sore throat but a painful head congestion. Why this different type cold from the same cold germ?

Or I may experience each and every cold in precisely the same fashion, with a burning sore throat followed by a general recovery or extremely brief, very mild additional symptoms. If a certain type of cold-causing virus causes this type of cold experience, why am I resistant to all other cold germs and non resistant to this one?

SPECULATIVE ANSWER:

A cold is experienced, as is any other bodily disease or lack of ease, in keeping with the pain of the spirit; the body responds where the spirit pains.

What spirit pain does the common cold express?

Rejection is almost always construed as an act of hostility, an act of non-love. We embrace what we love, reject what we fear, dislike, or for which we have no use. The body works within this general law. We evacuate our bowels, expelling matter that we don't need, for which we have no use. We urinate and perspire to rid ourselves of further waste. If we didn't rid ourselves of this waste material, we would soon die of our accumulated poisons.

When we are in the grip of a cold, we are busily expelling and ejecting, ie, rejecting.

A figurative expression of disdain—*I spit on you!*—tells us bluntly and succintly what we are doing when we have colds: *we are, quite literally, spitting on each other.* However politely we may try to do this, hastily bringing a tissue to our nostrils when we sneeze, covering our mouths when we cough, we can't entirely disguise what we are up to: *I spit on you, you spit on me.* This is what a cold is, a compulsive spitting on the world, an expulsion of liquid and not so liquid phlegm from the body. Explusion is rejection, an act of hostility, which means that the common cold is an expression of hostility, an expression of accumulated rage.

If the rage is there, it's a poison that must be gotten rid of. If the mind refuses the job, then the body takes over and comes down with a cold to sneeze, spit and cough up its anger and resentment. Once a thorough cleansing occurs, there is, most of the time, for most cold sufferers, a period of natural immunity before the buried resentments build up to discharge levels again.

Anger is a mental/emotional phenomenon, therefore the body expresses it on the mental/emotional level, primarily through the nose, which is the emotional connection on the mental level. The degree of chest involvement depends upon whether or not the anger goes that deep.

QUESTION:

What of the sore throat that so often precedes or accompanies a cold?

ANSWER:

When we draw in air, the breath of life, the flow is separated in our nostrils, but then reunited in our throats. Our throats express an underlying oneness/unity.

If we are feeling so hostile that we invite a cold in to help cleanse us, this is a violation of the oneness/unity expressed by the throat. Such a violation shames us so we burn with it, and we burn with shame right where the violation is occurring: in the throat.

QUESTION:

What of nose, head or chest congestion?

ANSWER:

Those of us living in this age, the latter part of the twentieth century, are extremely familiar with congestion. Traffic congestion plagues our highways and byways, often reducing traffic flow to a crawl.

Head—nasal—chest congestion is a similar stoppage of flow. Where there is an over abundance of accumulated rage, combined with a deep fear of expressing it—and if there weren't a deep fear of expressing it, it would never have piled up so abundantly—the flow of hostility is hindered, a bumper-to-bumper situation occurs, a bottling up that causes throbbing and aching:

Throbbing: "I want out!" "No, you don't!" "Yes, I do!" "No, you don't!"—the ache of fearful indecision.

If the hostilities run so deep that they have been packed down, buried, stored, out of sight and out of mind, congestion occurs in the chest, with a resultant coughing up of coagulated hostility: phlegm.

With all of these symptoms working in us and on us, we begin to feel we are dying and half wish we would.

But as the hours and days drag by, the poisons drain out, at their own rate of discharge, and soon we are feeling better, then in time we recover completely, feel well again, thank goodness.

Yes, thank goodness, thank the good that is in us that triumphs, quickly enough, over the stored up anger and fear. When suffering with a cold, we grow ashamed of our hostile feelings, cut off their expression and cure ourselves. As we grow older, if we also grow less angry, more loving, we cut down the incidence of colds. We can in fact become all but completely immune to them, experience an acquired immunity that has become natural for us. Colds become as much a part of our lost youth as adolescent acne.

If we feel anger/fear, these negatives must be dealt with—as any negatives/non-usables/poisons must be expelled from the body—but if catching a cold were the best way to deal with negative feelings, it would not be such an uncomfortable, distressful way.

On the other hand, of course, it's a reasonably safe way, for a cold is rarely to never fatal. Where an open verbal expression of all our stored up anger and resentment might bring numerous problems down on our heads, a cold ordinarily brings only physical distress. Failing to recognize the cold for what it is, an expression of hostility, or in any case not taking this hostility personally, those around us most often offer sympathy. Loved ones, in many cases, hover near, taking warm, considerate care of us. We have an excuse to stay in bed, if this can be arranged, an excuse to be waited on and to feel and act grumpy. Taken all in all, colds, no matter how miserable, are not all bad.

If you'd prefer not to have them, however, would prefer to banish the recurrent misery of colds from your life, all you need do is have your mind face and discharge the accumulated poisons so that your body doesn't have to. Possibly the current medical belief that the common cold cannot be conquered via a serum or serums is a true belief, for the present anyway, but anyone interested enough to work on it can banish colds from her or his life. All one need do is *tune in to the body, learn its language, and pay attention to what it says.*

Before we plunge into a detailed discussion of method, of how this switchover from body to mind can be achieved, let's have a look at another infectious disease, the once deadly plague, smallpox.

A typical case of smallpox manifests itself about like this:

Prior to the onset of symptoms there is an incubation period, believed to be from ten to fourteen days, then the disease strikes suddenly and severely. First there is a pronounced chill, soon followed by a fever. The temperature rises to 104 degrees Fahrenheit or even higher, while the sufferer experiences a quick pulse, intense headache, vomiting, and pain in the loins and back. These symptoms continue for three days, during the course of which time rash resembling scarlet fever or measles may appear on the lower part of the abdomen and inner sides of the thighs. On the third or fourth day following onset, the characteristic eruptions begin to appear, almost always on the face first, most often about the forehead and roots of the hair. The eruption spreads over the face, trunk and extremities, always most marked on the exposed parts of the body. On the second or third day, the original eruption, the papules ("*Pathol.* a small, somewhat pointed elevation of the skin, usually inflammatory but nonsuppurative") change into vesicles ("*Pathol.* a circumscribed elevation of the epidermis containing serous fluid; blister"). These vesicles show a slight central depression.

"The clear contents of these vesicles gradually become turbid, and by the eighth or ninth day they are changed into pustules ('*Pathol.* a small elevation of the skin containing pus.'). The fully developed pocks increase in size and lose the central depression."[1]

There may be marked "inflammation and swelling of the skin." Eruption may also be present on the mucous membranes, and the eyes may also be involved. With the appearance of the eruption, the temperature quickly drops, but in many cases there is a return of the fever as the vesicles are converted into pustules. This fever may be "accompanied by great restlessness, delirium or coma. On the eleventh or twelfth day the pustules begin to dry up and the febrile symptoms decline."[2] There is "great itching of the skin. . . The scabs produced by the dried pustules gradually fall off and reddish brown spots remain, which leave permanent white depressed scars—this 'pitting' so characteristic of smallpox being specially marked on the face. Convalescence . . . is as a rule uninterrupted."[3] If the victim has survived, the probability is great that he will now be immune to further attacks for life.

Let's take a close look at this appalling array of symptoms with reference to the three Good Health Hypotheses suggested earlier, particularly keeping the third one in mind: that bodily strain, with possible resultant damage, occurs precisely where the spirit experiences lack of a proper fit.

The onset of smallpox is sudden and severe, first rigor, soon followed by fever.

Rigor: "*Pathol.* a sudden coldness, as that preceding certain fevers; chill."

Our language tells us what emotion we are caught in when we are cold. We can feel "cold with fear," or "frozen with terror." Coldness can imply death—rigor mortis, the stiffening/coldness of death—unconsciousness—"out cold"—or sexual frigidity. To have "cold feet" is to lose courage, to be afraid. To be a "cold fish" is to be "withdrawn, dispassionate, one who displays no love, passion or affection."[4] We never say of someone that he is "cold and loving" anymore than we would describe someone as "warm and unloving." Warmth goes with loving, coldness with being unloving, and with being afraid.

Coldness/fear can be such a devastating emotion that to function at all we must push it away, and the energy with which we push it away is anger. Modern psychology has identified anger/fear as two faces of the same emotion. If we are frightened enough, we reach for the energy of anger to save us; if we are angry, underlying the anger is a fear we don't dare face.

Smallpox starts with fear, a fear that is soon pushed away by anger, ie, a fever.

Fever is an elevation of temperature; hotness.

Our language tells us what feeling hot means: "hot tempered," "hot under the collar," "hot and bothered"—all expressions linking being hot with feeling angry. There is another recurrent linkage as well, expressed picturesquely in the phrase "hot to trot."

"Most uses and meanings of 'hot' have a connotation of physical excitement tending toward, suggesting, or at least resembling sexual excitement."[5]

We can become aroused—worked up—heated up—either in anger or with sexual passion, *each being an alternative expression of the other.* A "fever" is an expression of hotness/anger/sexual passion.

Smallpox starts with fear, but the fear is soon pushed away by anger/sexual passion, the energetic emotion that saves us from being "frozen with fear," that is, saves us from total paralysis.

While in the grip of fever, the smallpox victim experiences "a quick pulse, intense headache, vomiting and pain in the loins and back," also in some cases a rash "on the lower part of the abdomen and inner sides of the thighs."

A quick pulse: When we're afraid or angry our hearts beat fast.

An intense headache: A headache feels as though ignorant armies were clashing on and in and through our brains, which is what a headache is: a clash/battle in the mind. With the onset of smallpox, the body is being forced to carry an extremely heavy burden and may not survive; shouldn't the mind open up and deal with the negatives itself? But the airy, avoidance prone mind would often rather die than face certain unpleasant truths about itself, so the mind darts away while the body, staggering under the horrendous load of negative messages registering on it, tries to avoid having to express them all, and the battle is on. The head hurts in keeping with the hell of a war being waged within it.

Vomiting: Food is the primary physical expression of love. When we vomit, we are rejecting food, nourishment, ie, love.

Pain in the loins: Anger is the other face of fear; sex is an alternative expression of anger. The anger the fever is expressing could be channeled into sex, and the loins ache with the hunger to do this. Any pain says: *pay attention; remember me!* The pain in the loins says: "Hey, you guys, you up there in my head fighting away, you who are so busy stirring up my blood to fever pitch, remember me: I'm your loins and can handle this for you, if only you'll let me!" But there is such a fierce war going on, this message gets lost.

Pain in the back: Man is distinguished from all other animals by the spine that keeps him upright. The pain in the back says: *"Pay attention: I'm an upright guy!"* Man is unique not only because he stands upright on two feet, but because he has a year round, rather than a periodically recurring, sex life. The upright stance—the spine—was instrumental in bringing this about and pain in the back should bring this to mind. The back pain reinforces the message of the loins "We could deal with this fever through sexual expression if only you dumb guys would pay attention!"

During this pre-eruptive stage of the disease, "there may occasionally be noticed, especially on the lower part of the abdomen and inner sides of the thighs, rashes resembling scarlet fever and measles."[6] Here the body is getting right down to cases, setting up a flashing red signal right around the accident scene, sending out a screaming reminder of how this entire problem could be handled—through the genitals—but who's listening? who's paying the least attention? Once war breaks out, who ever listens to reason?

Fortunately, after three or four days of this, the eruptions begin to pop out and the fever dies down. The heat is off—for the time being at least—as the conflict has been resolved. The body is now expressing the accumulated anger/sexual passion in a nonsexual way.

Skin eruption: a "non sexual" sexual release.

In smallpox the eruptions are first papules, a small, somewhat pointed elevation of the skin, usually inflammatory (reddish, swollen, warm, tender) but non suppurative (no production or discharge of pus), then they change to vesicles, (containing serous fluid), then with another change they become pustules (containing pus), at which time the smallpox victim may experience another fever, "accompanied by great restlessness, delirium or coma."

Compare this to the experience of sexual release:

With the first sexual arousal in the male, there is a "small, somewhat pointed elevation" of the penis, usually inflammatory (reddish, swollen, warm, tender) but non suppurative (no pus). With continued arousal, a clear, slippery, "serous" lubricating fluid appears, and with activity carried to climax the penis ejaculates, ie, ejects semen, a process often "accompanied by great restlessness (movement), delirium (of joy) or coma" (some people lose consciousness at orgasm, in earlier ages, "to die" was a euphemism for experiencing orgasm).

Pus: "a yellow-white more or less viscid substance produced by suppuration and found in abscesses, sores, etc., consisting of a liquid plasma in which leukocytes are suspended."

Semen: "the viscid, whitish fluid produced by the male reproductive organs, containing spermatozoa."

If there is the least substance in the contention that an eruption on the skin equates with the erection of the penis in sexual arousal, we will surely find corroboration for this apart from the pox of smallpox.

Let's have a quick look at acne.

Acne: "*Pathol.* an inflammatory disease of the sebaceous glands, characterized by an often pustular eruption of the skin, esp. of the face."

Acne tends to be a plague of adolescence, blighting those teenage years between puberty and adulthood. The body matures sexually at puberty to where it becomes capable of reproducing itself (the definition of puberty), which in common law is considered to be fourteen years in the male, twelve in the female. Oftentimes, while the body may be ready at these early years, the thinking/feeling human being inside the body has a ways to go, and sexual frustration may occur. Result: adolescent acne, pus-filled eruptions carrying the burden of sexual expression.

Years ago I was delighted to hear Orson Bean, when asked on a TV game show whether he still remembered his first lover, affectionately mention a girl's name and then remark with gratitude that she had "saved him a fortune in Clearasil." When I mentioned this to a Chicana friend, she laughed and said that it is commonly accepted folklore among the Mexican people that acne is an expression of sexual frustration. Adult males tease pimpled teenage boys, she said, that all they need is a good woman to clear their skins.

SPECULATIVE SUMMARY:

Smallpox could be considered a dangerous, sometimes lethal form of acne, or acne a mild form of smallpox. Both are poxes upon our physical houses.

QUESTION:

But if the above has the least validity, why would anyone in his right mind chose the misery, death defying risk and almost sure disfiguration of smallpox over the generally pleasant experience of sexual intercourse?

ANSWER:

Why did so many early Christians chose martyrdom, which at first glance seems a similar, irrational choice?

If we look more closely, possibly we will see that it isn't entirely irrational, after all.

If one sincerely believes that life on earth is brief and matters only as a preparation for eternal bliss, it is not too difficult to understand why such a sincere believer would cling to his faith, and to his virtue, even in the face of sure death.

Smallpox raged as a plague during those centuries when the Christian Church, with its message of the sinfulness of sex and the glories of eternal life which awaited the righteous, reigned supreme.

For a sample of the Church's attitude on sex, consider the essay "Of the Honorableness of the Marriage Bed," by St. Francis de Sales (1567-1622), there being, of course, no such thing as "honorable sex" apart from the "marriage bed."

> "'The Marriage Bed' must be 'undefiled'—that is to say, exempt from immodesty and other defilements. Thus was marriage first instituted in the

> Earthly paradise where, until the time of the fall, there was no disorder of concupiscence in it, nor anything dishonorable.
>
> "There is a certain resemblance between the pleasures of the flesh and those of eating, for both of them relate to the flesh, although the former, by reason of their brutish vehemence, are alone called carnal. I will therefore explain what I cannot say of the former, by what I shall say of the later."

St. Francis, in discussing the "honorableness" of marital sex, has already cast several severe aspersions upon it. Marriage as first instituted, ie, prior to the time sex reared up in all its ugliness, was obviously to be preferred, for then it had nothing "dishonorable" in it. Concupiscence is a "disorder," the pleasures of the flesh have a "brutish vehemence." With this kind of defense, what is left for any who might wish to launch an attack?

St. Francis continues by pointing out that just as food is necessary for the preservation of life, sex is necessary for the procreation of children, therefore it is "good, holy and commanded." One spouse is not to withhold sex from the other, but is to allow it faithfully, freely, with good spirit—as long as sex is not indulged in "to excess and immoderately. . .

> "Of a truth, nuptial intercourse which is so holy, so just, so commendable, so useful in the commonwealth, is nevertheless in some cases dangerous to those who make use of it: for sometimes it makes their souls very sick with venial sin, as happens by simple excess; and sometimes it causes them to die of mortal sin, as happens when the order established for the procreation of children is violated and perverted; in which case, such sins are always mortal. . ."

Continuing with the analogy he has set up—eating as similar fleshly pleasure standing in for sexual indulgence—St. Francis writes:

"It is a true mark of a beggarly, mean, abject and base spirit to think of the dishes and of eating before the time of the repast, and more so still, when afterwards one is taken up with the pleasure which one has had in the meal, dwelling upon it in words and thoughts, and allowing one's mind to wallow in the remembrance of the pleasure enjoyed in swallowing down the mouthfuls; as do those who before dinner have their mind fastened on the spit, and after dinner on the dishes; persons fit to be scullions. . . Persons of honor do not think of the table until they sit down to it, and afterwards they wash their hands and mouth, in order to lose both the taste and the smell of what they have eaten."

St. Francis then explains how this "honorable" discord of sex should be handled:

> "The elephant is only a huge animal, but he is the most worthy beast that lives on the earth, and the most intelligent. I will give you an instance of his excellence; he never changes his mate and tenderly loves the one of his choice, with whom nevertheless he mates but every third year, and then for five days only and so secretly that he has never been seen to do so; but he is seen again on the sixth day, on which, before doing anything else, he goes straight to some river, wherein he bathes his whole body, for he has no wish to return to the herd until he has purified himself. Are not these excellent and modest traits in such a beast, by which he invites married persons not to allow their affections to remain attached to the pleasures of sense which they have experienced in accordance with their state of life, but, when these are past, to wash their heart and affection of them, and to purify themselves of them as soon as possible, that afterwards they may perform other actions which are more spiritual and lofty?"

This sets it up very clearly, for the devoutly believing Christian:

If one strives for excellence, sex should be indulged in but once every three years; one should then purify oneself immediately afterwards, never think about it beforehand nor allow the mind to enjoy the memory of it afterward. Anything beyond this is excess and excess is dangerous: it will sicken the soul. Sex was "established for the procreation of children" and if this natural "order" is "violated or perverted," it is always a mortal sin, and the sinner will be damned.

Who could be sure of living up to these strictures, which covered not only the area of action but extended into the mind as well? Who could keep all stray thoughts of sex away, could keep himself from anticipating beforehand or enjoying in memory afterwards? Yet such lapses were venial sins which sickened the soul. To go even further, to dare to pervert sex, that brought nothing less than eternal damnation. To lose one's life was a relatively minor thing, but to burn forever in the fires of hell, forever and ever, time without end—No wonder the pox looked almost benign in comparison.

It is interesting to speculate on whether, or how often, the plague flared up coincident upon the emergence of some eloquent new preacher with a fiery tongue able to preach mighty sermons of warning to the eager faithful, who trembled in their minds and hearts as they listened—and crawled home chastised, to be stricken within days with the chills and fever which signaled the pox.

It is not suggested that only sinners were smitten with smallpox, or that the worse the sinner the worse the pox. The reverse was more likely true: the more delicate the conscience, the more piercing the sense of sin, the greater the susceptibility to smallpox. After all, it all depended on what one feared. All the pox could do was kill you, while being disorderly, dishonorable or perverted in your sexual thoughts or actions could damn you to an everlasting hell. Little wonder that in those sin-ridden centuries, the plague flourished as it did.

With the industrial revolution and the age of enlightenment leading into the modern age, smallpox was effectively conquered, along with the deep, dark, all pervading sense of sexual sin.

QUESTION:

But if smallpox results from an individual spirit-pain, from a psychic disturbance/ill fit on the part of the individual, how can we explain the epidemics when masses of people were afflicted?

ANSWER:

The same way lynchings, riots, wars or any other irrational mass action is explained: through man's high degree of suggestibility. If most people suffer, at any given time, from the same repressed angers/fears, when a few start expressing these through bodily disorder, it is easy for everyone susceptible to jump into bed to follow the leaders. Hysteria can sweep through a crowd, causing a riot. Throughout history young men have been handed weapons and exhorted to go kill, Kill, KILL! Caught up in war fever, off the young men go, to indulge in the dangerous, often lethal disease of war. Whether the fever is based on racism, religious belief, territoriality, or nationalism, no one suggests that this plague, this irrational mass disease, is caused by a physical bug, a microbe or virus that can be isolated in the laboratory and observed.

The same irrational suggestibility which causes wars can operate with infectious diseases, bringing plagues.

SPECULATIVE CONCLUSIONS:

The common cold is an expression of repressed anger/sexual feeling, the two passions being basically interchangeable.

Smallpox is an expression of anger/sexual feeling that could be expressed sexually but isn't.

From the above, the extrapolation:

HYPOTHESIS:

All infectious diseases are caused by repressed anger/fear.

****

Before we proceed to a discussion of non-infectious diseases, let's take a very brief look at one other infectious disease, another old time scourge, yellow fever.

Yellow fever "first traveled west on slave ships in the seventeenth century." It "rampaged the eastern and southern cities of the United States." It killed "quickly and horribly." The victims turned "lemon yellow and the terrifying black vomit started."

Dr. Benjamin Rush, "the most eminent physician in Philadelphia," attended patients from dawn till dark in the yellow fever epidemic which struck his home city in 1793.

> "On entering a patient's room, the first thing that struck Rush was the victim's countenance. 'It was as much unlike that which is exhibited in the common bilious fever as the face of a wild is unlike the face of a mild domestic animal. The eyes were sad, watery, and so inflamed in some cases as to resemble two balls of fire. The face was suffused with blood, or of a dusky colour, and the whole countenance was downcast and clouded . . . sighing attended in almost every case.'"[7]

Let's take a close look at these symptoms.

Sighing: We often sigh when we feel depressed or deeply regretful.

Eyes sad and watery, ie, tearful: Tears can be caused by grief, regret, fear, or by positive emotions such as joy, but here the eyes are "sad" as well as tearful, which would rule out joy. The eyes are also in some cases inflamed with blood, ie, anger/passion. Accompanied by sadness, this red blood/anger/passion would seem to be turned inward upon the self, not outward upon the world.

The face was suffused with blood: We can be red-faced with either anger or shame, but in this case, with the "whole countenance . . . downcast and clouded," the face would seem to be blood-red more in shame than in anger.

The look of the face suggested to the good Dr. Rush the difference between the face of a "wild" as distinguished from a "mild domestic animal."

Yellow fever journeyed west with the slave ships. On these ships were black men, believed to be "wild," uncivilized, not quite human compared to the civilized, "mild domestic" white men who enslaved them.

Yellow fever ravaged the "eastern and southern coasts," that is, all the port cities, all or most of which were active in the slave trade.

Our founding fathers wrote in the Declaration of Independence that "all men are created equal," but this use of the term "men" did not include women, who were not considered equal, nor did it include those males with black skins, who were also not considered equal. To be female or black was not only to be something less than equal, it was also, at least in the case of the blacks, to be considered something less than human. As late as 1900 Christian preachers in this country were still giving sermons to "prove" that "niggers have no souls."

This denigration of the black race into something less than equal, something not quite human, had to be accepted or how could one be both a good Christian and a dealer/trader/owner of slaves?

But what if—what if—the suspicion began to creep in that slaves weren't "wild" animal creatures after all, but were instead fully human, people who had every right to be treated as equals, with dignity? How did a slaveowning population deal with that, especially in the port cities where the importation and sale of slaves flourished right under its nose?

If this terrifying awareness of the humanity of black people couldn't break through into the mind, it *could* break through to color the skin, color it with the yellow of awareness, causing the eyes to become sad and watery, also in some cases inflamed with anger directed inward. It could suffuse the face with shame—and then there was the "terrifying black vomit," vomit colored to show that inside the white skin lived the same blackness, the same humanity—or inhumanity—that belonged to the slave. Under the skin, all were alike, all were brothers. If black people brought in chains from their home continent into slavery in a faraway country were indeed human and equal, then the white Christians enslaving them had better beware of the peril they were in, the eternal damnation they faced if they didn't open their eyes to face the enormity of what they were doing. But to become aware—oh, sweet Jesus, who had repeatedly commanded us to love one another, to love our neighbors as ourselves, how could the mind accept such a staggering guilt? Let the body carry the burden, express the awareness and the pain, even if the fever proved fatal. Better that than to have the mind crack under the strain of such an insupportable weight.

Those in Cuba who volunteered to be infected with yellow fever in order to help in its eradication had an astonishing recovery rate. One can assume that they felt far less crippled by guilt than those who fell victim to the disease under less humanitarian circumstances, therefore their disease proved milder and non lethal.

With a general recognition by the white Christian world that the enslavement of their black brothers was an evil not to be condoned ever again, yellow fever was conquered—throughout that part of the world inhabited by the white skinned race. It has since reappeared in other areas, but that is another story and not one to be dealt with here.

SPECULATIVE COMMENT:

Susceptibility to yellow fever was caused by guilt/fear, guilt being simply another name for fear.

To recapitulate:

HYPOTHESIS:

All infectious diseases are caused by the negative emotion anger/fear.

Note: I did not choose the three infectious diseases discussed—the common cold, smallpox and yellow fever—due to any belief that they would most easily fit into a preconceived framework, but for the following reasons:

The common cold. We are almost all familiar with the common cold and know from experience that it is a communicable disease, that it can sweep through a family from one member to another. This would seem to be close to absolute proof that it is, as the medical profession claims, caused by a germ. Brought up to believe in the modern germ theory of disease, we tend to forget that other things are also infectious: smiles, laughter, yawning, hysteria, almost any strong emotion such as anger, fear, et cetera. No one concludes that because these are communicable from one person to others that they are also caused by germs.

I wanted to start with something with which there would be a general familiarity, also a disease commonly believed to be transmitted by germs, and nothing seemed to fit these two criteria as well as the common cold.

Smallpox. This was, as far as I knew, the first infectious disease generally believed to be wiped out by modern scientific methods, which is why I chose it.

Yellow fever. In the marvelous book, *The Virus That Ate Cannibals, Six Great Medical Detective Stories,* by Carol Eron, in which the author details the efforts of modern medicine to track down the cause of and eradicate different infectious diseases, the first disease Eron deals with is yellow fever. I chose it for this reason, because it was the subject of the first "great medical detective story" in her book.

Let's turn now to a discussion of non infectious diseases to see if they fall into any observable pattern.

8

BODY-SQUEAK

If Our Bodies Ran on Wheels,
Would We Heed the Squeaking/Squealing?

Fate is nonawareness.
Jan Kott

It is generally recognized that the average American tends to take better care of his automobile than he does of his body.

A car is ordinarily taken in for regular tune-ups/check-ups, if the owner can possibly afford it. Pings and squeals are listened to, every attempt is made to keep the machine running at maximum efficiency. Few car owners, in response to a warning signal, would rely on lifting the hood and tossing a couple of pills down the radiator, secure in the feeling that within a few minutes relief will arrive and all problems will magically go away.

This mystical cure-rite is reserved for the physical body, one's car being far too important to be treated in this slap-happy fashion.

There is today such a wide array of medications to alleviate human ills as to stagger the imagination. No matter what area of the body hurts, there is a pill to assure relief.

If you have a headache, pop a pill.

A stomachache? Pop a pill.

A backache? Here, down two of these.

Eyestrain, earaches, discomfort in your nose, throat or sinuses, cold sores on or in your mouth, a cough, pain in any part of your torso or limbs, don't worry about it, don't feel concerned, your kindly, prescription-writing doctor or friendly neighborhood druggist can give you something to keep you from hearing your body's squeak.

If you're feeling anxious, frightened, lonely or depressed, cheer up, there are pills for that too, tranquilizers to soothe and comfort you, to make the squeal go away. In this day and age, no one has to listen to the warning signals his body sends out. We can muffle them, stifle them, chemically knock them out, secure in our faith that this mystical cure-rite will not only bury the symptom, but will somehow, through magic, kill off the cause of the symptom as well.

SPECULATIVE COMMENT:

To feel fit, we must have a proper mind/spirit fit.

To know whether we have a proper fit, we must listen to the body's squeaking/squealing.

As with various automobile pings and hisses, we can deal with a minor problem now or ignore it, except for popping pills, and deal with a major breakdown later. The choice is ours.

****

Modern medicine has divided physical diseases into two broad classifications: infectious and non infectious. We have dealt in the preceding chapter with a sampling of the former. Let's now have a look at some typical cases of the latter.

Just as the common cold is probably the most widely experienced of those diseases considered infectious, there is a physical disorder so commonly experienced in our midst today that almost everyone has some acquaintance with it: the need for eyeglasses to aid sight. Though we ourselves may have escaped this plague, we almost certainly know someone who is nearsighted, farsighted, or suffers from astigmatism, eye fatigue or some other condition which requires the periodic or continual use of glasses to boost or clarify his vision.

Let's take a closer look at an extremely common vision problem: myopia or nearsightedness.

Those afflicted with this condition can see clearly close up, but objects at a distance are blurred. For objects to be seen clearly the image "must be focused on the screen of nerve endings (the retina) at the back of the eye. In myopia, objects at a distance are not focused on the retina but at some point in front of it."[1]

There are several explanations offered for this error of refraction.

One explanation offered—the one most widely believed—is that the eye is too long, from front to back, causing light rays from distant objects to focus short of the retina. Or an eye of normal size can become myopic, it is believed by some, due to an increased curvature of the cornea and lens, causing too great a refraction (bending of light), resulting in the same shortened focus.[2]

There is a school of thought that believes that myopia is caused not by the inherited shape of the eye but is brought on by eyestrain, from too much reading or other close work done under too poor a light.

Nutritionists advance the theory that myopia is a "disease of the connective tissue" of the eye. "In 1956 another researcher, Garzino, using an electron microscope, discovered that the fibers of collagen (that is, the connective tissues) in the eye of a myopic individual have a smaller diameter than those of the normal eye and that they are bathed in a more abundant liquid than is present in the normal eye."[3] The collagen fibers of the eye "may lose part of their physical qualities and become abnormally distended so that they cannot give the support they should. In other words, the ligaments become lax."[4]

Another researcher, Dr. W. H. Bates, an ophthalmologist who studied the problem of myopia and other vision problems through long years of experiment and clinical practice, concluded that nearsightedness is the result of strain leading to poor visual habits.

The mode of dealing with myopia varies with the belief as to its cause. The most widely held view is that myopia is an inherited condition. "There is no evidence that reading (even in poor light) or the use of the eyes in any way will influence myopia, or that it can be changed by doing 'eye exercises' or by wearing glasses. The progress of myopia can be neither halted nor accelerated until, like one's height, growth ceases and full maturity is reached."[5] Instead of pills to pop to correct the condition, glasses with concave lenses are prescribed. Just as most pills today provide relief from the symptoms without effecting any cure, "It should be emphasized that eyeglasses are worn to provide normal vision and not as a treatment for myopia,"[6] for which there is no cure.[7] A crutch, yes—eyeglasses or contact lenses—but no cure.

Those who ascribe to the nutritionist theory of myopia claim that the widely accepted view outlined above is in error and that it *is* possible to halt the progress of the condition. "Drs. C. and G. Desusclade tell us that they have been giving vitamin E for the past 10 years and that they can generally stop the process of the disease, or maladjustment so that the young patient does not become any more nearsighted. Checking on their early patients after 8 years, they found that the effects are lasting."[8] What causes myopia? "'Lack of vitamin E or perhaps an exaggerated need for this vitamin or a lack of ability to assimilate the vitamin properly,' they declare. They continue, 'Proof of the vitamin E treatment lets us verify not only the complete stoppage of the development of a case of myopia, but also the improvement of those little, more general symptoms that are so often associated with it—little disorders of the joints, slight glandular upsets, retarded development of puberty.'"[9]

Dr. W. H. Bates, who considered myopia as due to strain and poor visual habits, claimed repeated cures with his method of relaxation techniques and visual reeducation. Visual accommodation—the ability of the eye to focus first on a near object, then on one farther away—is accomplished, Dr. Bates contended, by the external muscles of the eyeball. These muscles are functioning improperly if myopia is experienced. So why do these muscles so often fail to function properly?

"It has been demonstrated in thousands of cases that all abnormal action of the external muscles of the eyeball is accompanied by a strain or effort to see, and that with the relief of this strain the action of the muscles becomes normal and all errors of refraction disappear. The eye may be blind, it may be suffering from atrophy of the optic nerve, from cataract, or disease of the retina; but so long as it does not try to see, the external muscles act normally and there is no error of refraction. This fact furnishes us with the means by which all these conditions, so long held to be incurable, may be cured."[10]

In this myopic age, many of us have preconceived notions as to what type of person wears glasses. If we were asked to visualize first a college professor and then a street mugger, we would quite likely put glasses on the former and not on the latter. Various studies have shown that such stereotypical thinking, quite possibly derived from experience, has more than a little validity. Myopes *do* tend to share certain traits that distinguish them from non myopes.

Psychological testing of 140 male myopes and 118 male non myopes belonging to the entering class at a military academy showed that the myopes scored higher on academic achievement and creative performance while the non myopes were more oriented toward business, selling skills and outdoor activities. The myopes were "significantly more likely to achieve officer status and win more academic awards than the non myopes."[11]

In another study, Raymond M. Singer, Washington State University, researched a possible relationship between visual refractive error and various personality factors and academic performance. In earlier studies, Mr. Singer noted, myopes had been described as introverted, more interested in ideas than in action, and academically oriented.[12]

For his own research, aimed at delineating any relationship between psychological variables and myopia, Mr. Singer worked with approximately 300 male and female university students. The test subjects were examined visually and given psychological surveys. Grade point averages and pre-college aptitude test scores were obtained. The results?

"The myopes scored higher on practically every measure of academic aptitude, also achieved higher college grade point averages. Myopes prefer mental to physical activity and prefer to work independently. Myopes value *wisdom* and *independent* more than do the non myopes, and value *clean* and *helpful* less."[13]

Myopes scored lower on the degree of super-ego (in Freudian terms, the conscience) and on proneness to guilt: "They respond more rationally to situations where others might feel guilty." Compared to the median test on the total sample, myopes were found to care more for the *intellectual* and for *mature love,* and to care less for "a *comfortable life, honest* and *an exciting life.*"[14]

Mr. Singer concluded: "These results suggest it is possible to classify individuals as to myopia before the condition develops, so that the person could be referred to an eye physician for preventive treatment."

To suggest that once pre-myopes are identified they should be referred to an eye physician for preventive treatment seems a rather curious conclusion to draw from the research results that precede it.

Ophthalmologists—eye physicians—generally accept the idea that myopia is an inherited condition, that properly fitted glasses can correct the myope's vision but can do nothing to cure the condition or halt its progressive nature. Simple myopia "is rare at birth but usually begins around 6 or 8 years (a bit earlier for girls) and increases as the body grows."[15] Even if all future myopes could be identified through psychological testing at age four, a couple of years before myopia makes itself evident, what could the eye physicians possibly do to help? All they can do is prescribe glasses, and to put the pre-myopic child in glasses before they are needed would be like putting someone on crutches before and not after he breaks his leg.

In Mr. Singer's study, it was found that there is a *psychological* difference between myopes and non myopes.

Therefore if you classify individuals prior to the onset of myopia on the basis of their psychological characteristics, in order to send them for preventive treatment, surely it is to a *psychologist* that they should be sent. According to Mr. Singer's study, if you could subject the pre-myopes to psychological conditioning to where they would learn to value wisdom and independence less, value being clean and helpful more, could also inject them with greater conscience and a higher level of guilt, teach them to put less value on the intellect and on mature love, if all this could be accomplished, the myopes would no longer differ from the non myopes and possibly through this psychological reconditioning the onslaught of myopia could be averted.

Also if we could catch children young enough and make sure that none became introverted, that all grew up to be more interested in action than in the world of ideas, and above all made sure that none became oriented toward academic achievement, we might possibly wipe out the myopic eye for good and all.

Another study on "Personality Factors Related to Development of Myopia," in which subjects were given various personality tests, found that: "Moderately myopic subjects were found to be the most highly differentiated from emmetropes (those with normal vision), and could be described as seeing the world as noxious, irrational, and pressing on them."[16] This intriguing finding tends to suggest that myopes blind themselves to the world "out there" because they find it unbearably ugly and too close for comfort. The treatment offered in connection with the study: relaxation and guided fantasy.

There is a school of thought which maintains that the myopic eye is a failure of evolution. For eons man, the hunter, needed distant not near vision and his eye, adjusted for distant vision at rest, perfectly suited his needs. But with the advent of mass literacy, newspapers, books, movies, and that utterly delightful moving and talking screen, the television, the demands made upon the eye for near-seeing have simply overwhelmed it. The unnatural strain of forcing accommodation to all these near objects has been more than the eye could handle, and widespread myopia is the result. We can have our modern print culture and poor eyesight, or can return to a more primitive way of life and once more enjoy the normal vision still the fashion among less civilized people. In this view, normal vision has been sacrificed upon the altar of civilization and the myopic eye is the eye of the future, for all of us.

Dr. W. H. Bates, who spent his adult life studying visual problems, considered this nonsense. The creative force which fashioned us did not stick us with a defective, non adaptive eye. Primitive people did and do a great deal of near work. The women are seamstresses, embroiderers, weavers, artists creating "all sorts of fine and beautiful work. . . Yet women living under primitive conditions have just as good eyesight as the men."[17] It is not because modern, civilized man does a great deal of near work that he loses normal vision, according to Dr. Bates, but because, straining to see, he falls into poor visual habits and loses the natural art of seeing. Relaxation, reeducation and eye exercises will correct all errors of refraction, Dr. Bates insists.

The myopic eye was found to be "greedy." Too anxious to see everything all at once, it has lost the art of shifting the way a good eye should. With normal vision, objects are visually recorded portion by portion a bit at a time. The myopic eye is in too big a hurry for this and attempts to take in the entire forest all at once, unmindful of the individual trees. Myopes also tend to stare too intently, dimming their vision as a result. In addition, the myopic eye is too dry; it fails to blink and tear sufficiently. The eye with normal vision sees one part of what it looks at best, and all other parts less clearly. This is what Dr. Bates calls *central fixation,* but the myopic eye, operating under strain, can no longer see best the point at which it is looking—it has lost the relaxed art of seeing through central fixation.

Once the myope has fallen into these bad visual habits, he compounds the problem by grabbing immediately for crutches—he gets fitted for glasses—which put an immediate end to any hopes for improvement and ensure that his vision will deteriorate even more. On the dedication page of Dr. Bates' book, *The Cure of Imperfect Sight by Treatment Without Glasses* (retitled in later editions *Better Eyesight Without Glasses*), is the following:

> *"On a tomb in the Church of Santa Maria Maggiore in Florence was found an inscription which read: 'Here lies Salvino degli Armati, Inventor of Spectacles. May God pardon him his sins.'*
>
> *"Nuova Enciclopedia Italiana, Sixth Edition."*

SPECULATIVE COMMENT:

With vision problems, as with all other breakdowns in bodily functioning, the medium is the message.

The eye exists entirely on the mental level and connects directly with the brain.

According to Dr. Bates' long, intensive study, the myopic eye has lost central fixation, is greedy, stares too intently, and is too dry, failing to blink and tear enough.

Central fixation: The central feature of anyone's life is his own being, primarily his feeling self for above all else we are creatures dependent upon our hearts. Myopes have lost the natural, easy fixation on themselves. This is verified by studies which have shown that myopes are primarily interested in the world of ideas, that academically they are high achievers, that the myopes' approach to life is intellectual, not emotional.

Greediness and staring: The myope's intellectual approach to life includes the drive to take everything in at once, to stare down the world, to see life steadily and see it whole. How else can one achieve understanding and control? And without understanding and control, how can one hope to attain wisdom and independence, traits highly valued by the myope?

Failure to blink and tear: To blink is to lose sight of, and when the world is seen not as a beautiful, restful place but as noxious and irrational, who would dare close his eyes, even momentarily, and lose sight of the noxious irrationality pressing in on him? Eyes that tear are eyes that cry, and to cry is to be emotional and centrally fixated in a way that the myope has forsworn.

When we are walking in the dark with the aid of a flashlight, we illuminate the path at our feet or those other areas which might threaten our safe forward progress. When we drive our automobiles at night, we turn on our headlights. Often when we attend a stage presentation, a spotlight directs our attention to the performer we are meant to watch. These visual aids—flashlights, car lights, spotlights—though lighting up narrowly restricted areas, nevertheless give us the help we need. With these aids, we are in much the same situation as the myope, who can see clearly and distinctly close up but for whom distant objects are fuzzy and blurred.

HYPOTHESIS:

The message of the myopic eye is this:

If you want to see life steadily and see it whole, if you want to understand it, control it, become wise, gain independence, forget what's "out there," at a distance, which is of little or no importance to you. Study that which is near to hand, look inward, *know yourself.*

Only because this message is not sufficiently absorbed by the conscious mind, and is insufficiently acted upon, does the eye become nearsighted to start with and ever more nearsighted as the guiding spotlight contracts into a closer and closer range, doing its best to drum the message in: If you would know the world, know it steadily and know it whole, *first of all know yourself.* The myopic eye can relax into normality, can regain normal distance vision, once this message gets through and is acted upon, just as the flashlight and car lights and spotlights can be switched off once such visual aids are no longer needed.

****

Let's take a quick look at another visual problem, one that's the reverse of myopia: hypermetropia or hyperopia, far sightedness.

Whereas approximately one fifth of the adult population now experiences some degree of myopia, hyperopia is even more widespread, is in fact the most common of all problems of refraction. Estimates are that it affects up to two out of three of all adults. Like myopia, it is believed, by the medical profession, to be caused by the shape of the eye. In hyperopia the eye is too short, from front to back, or, far more rarely, the curvature of the cornea is insufficient. In either case, the image picked up by the eye is not focused properly on the retina but comes to a focus behind the retina. Convex lenses correct the error of refraction.

Whereas the myope rarely to never experiences any symptom of eye strain—distant objects are fuzzy but there is no problem in viewing them—those afflicted by the opposite condition, farsightedness, commonly experience headaches and eyestrain when they are forced to use near vision.

SPECULATIVE COMMENT:

Here again, as with myopia—as with any bodily malfunction—the medium is the message.

The clear distance vision is a spotlight directing the attention of those afflicted.

HYPOTHESIS:

The message of hyperopia is clear:

It causes strain and pain to be so wrapped up in the near, so stop paying so much attention to every petty little object and happening close to hand. Divorce yourself from the near; gain distance and perspective. Stop concentrating on every bug on every little tree and learn to view the forest.

Here again, as with the myopic eye, it is only because this message is not sufficiently absorbed by the conscious mind, and is insufficiently acted upon, that the farsighted eye becomes and remains farsighted. Once a sufficiently distant, detached view of the world is achieved, so that every little happening no longer causes strain and pain, once this message gets through and is acted upon, the hyperopic eye can relax into normal close vision, just as any visual aid can be switched off the moment it's no longer needed.

****

Leaving eye problems behind, let's take a look at another non infectious condition, another extremely common affliction, one that causes an immeasurable amount of pain and anguish: arthritis.

Arthritis: "Pathol. inflammation of a joint, as in gout or rheumatism."

The term "arthritis" refers not to a single disease but rather to "a related group of more than a 100 different conditions."[18] According to the Arthritis Foundation, about 36 million Americans suffer from some form of arthritis. One in three families is affected, with a million new victims stricken each year. The disease is older than man, and has plagued mankind for at least 100,000 years. "Subsequently another skeleton of Neanderthal man was found, with the spine almost intact. . . a careful study of the. . . find revealed that the spine was deformed by osteoarthritis."[19]

With this long history of familiarity with the disease and its widespread occurrence, how much do we now know about the cause, treatment and cure?

Not very much.

> "Sometimes arthritis strikes suddenly, only to fade away and disappear without a trace. At other times its progression is inexorable, causing unremitting suffering daily. It is a disease often marked by sudden, unexplained periods of remission that end as unpredictably and mysteriously as they began. Moreover, treatment that brings help and relief to one person may prove to be of little value to another.
>
> "All of which is to say that the cause of most forms of arthritis is still unknown and that the means of prevention or cure—except for a very few kinds—have yet to be found."[20]

Eighty-five percent of those who suffer from some form of arthritis are victims of either osteoarthritis or rheumatoid arthritis, with gout trailing behind in third place.[21]

Let's take a closer look first at rheumatoid arthritis, "The great crippler."

"Rheumatoid arthritis is a mystifying disease. . . RA follows its own special pattern and no two cases are quite the same."[22] The "cause of RA is still a mystery, and treatment, to a considerable extent, a matter of trial and error. . ."[23]

Like most forms of arthritis, "RA affects some parts of the body far more frequently than other parts. . ."[24] The joints most often affected are those in the fingers and toes, followed by the wrists, knees, shoulders, legs, elbows, ankles, neck and jaw. "The hands may become virtually unusable, and the feet so distorted that special shoes are needed. The disease can produce such inflammatory changes in connective tissue that tendons break so that their muscles no longer connect to bones and these bones can no longer be moved when the victim wants or needs to do so."[25]

The disease can develop suddenly or slowly. It may appear in mild form, soon fading away, or flare-ups may alternate with remissions. "Days or weeks may pass between these flare-ups and remissions, but gradually the disease clings to its victim for longer and longer periods until it is present virtually all the time."[26] No doctor can predict whether any given case will be very mild or totally crippling.[27]

The cause of the disease is not known but the medical profession has put forth three leading theories:

1. The infectious theory, though no causative organism has ever been found.

2. RA is caused by a malfunction of the body's immune system; the system goes haywire and the body attacks itself.

3. A combination of the first two: infection triggers malfunction of the immune system.[28]

"Psychological stress has also been suspected but finally exonerated as a cause."[29]

The Arthritis Foundation reports that the onset of the disease usually occurs in those from twenty to fifty years, but it can strike children too and also the elderly. Two-thirds of the sufferers are women. The disease affects the entire body. There is fever, weariness, loss of appetite, anemia; the lungs may be affected with pleurisy developing. "RA may affect the skin or blood vessels, the heart or eyes or muscles in a variety of ways. Sufferers are often bothered by cold, tingling hands and feet or excessive sweating. There may be lumps under the skin. . . Typically not one but many joints are involved and become inflamed."[30]

Treatment consists of a "total therapeutic strategy."[31] This includes rest, physical aids such as splints or crutches, physical therapy such as heat, exercises, drugs, possibly surgery, and "very likely a new and long term life-style."[32] Aspirin is the "backbone and mainstay of all arthritic medication," but numerous other drugs are also being tried, many of them potentially dangerous. Gold injections "can cause improvement and even remissions in about three-quarters of those able to continue taking them."[33] Possibly the brightest new hope is surgery, wherein the damaged old parts are yanked out to be replaced by new. However, this is primarily "a court of last resort, when pain, stiffness, disability, and deformity are such that the sufferer can no longer live with the condition."[34] Surgery is also "essentially a confession of failure—a failure of medical science to solve the mystery of the disease so it can either be prevented, adequately controlled or cured. Nor is surgery totally curative; it can only relieve a particular set of local effects or systems, not cure the arthritic condition."[35]

Medical science currently rejects any notion that diet or climate can play any significant part in controlling or alleviating RA.

Nutritionists, despite the denials of medical science, claim significant success treating RA through dietary means, with Vitamin E therapy proving, through experiment, to be extremely beneficial.[36] In another experiment, twelve hospitalized arthritis patients were put on a raw food diet, with no other treatment given. "Eight of the patients began to feel better in one to four weeks on this diet. Two improved up to 5 or 6 weeks, then relapsed. Two showed no improvement at all. In the follow-up, after the patients who improved had gone home, it was found that 7 of them continued to improve to a marked degree."[37]

Various studies have been done on the psychological characteristics of people who suffer from rheumatoid arthritis. In one study "Cleveland and Fisher found that in interviewing arthritics, they appeared overtly to be calm, composed and optimistic individuals who rarely if ever express or even feel hostility. When hostility is expressed it goes inward. They complain of their own social shyness and social inadequacy."[38]

Another investigator, Southworth, "postules the theory that the arthritic process could begin with conscious aggressive impulses that become intolerable. If the 'hard shell' of muscle tension was once consciously erected against the disturbing aggression, and then the whole conflict was repressed, we have inhibited aggression as a prominent feature in an essentially hysterically initiated illness. In a comparison of arthritic patients with ulcer patients, arthritics tend to be more disturbed in their relationships with their parents than ulcer patients."[39]

Another researcher, Stanley King, "delineated what he thought were the common factors of the personalities of rheumatoid arthritics. They:

"1. Lead quiet lives;
"2. Complain of shyness and inadequacy;
"3. Have marked feelings of inadequacy and inferiority;
"4. Are self-sacrificing;
"5. Are overconscientious;
"6. Have a strong need to serve others;
"7. Have obsessive-compulsive characteristics;
"8. Have a tendency to depression.

"The most prominent problem is that of aggression. Both children and adults are described as having a strong reservoir of hostility below consciousness, primitive in nature with sadistic and destructive fantasies. However, the arthritic is fearful of his anger and resentments, fearful that expression of these feelings will have drastic consequences for him.

This may account for his emotional self-restriction and his inability to show anger. . ."[40]

In *The Psychological Aspects of Rheumatoid Arthritis,* the author, Harold Geist, Ph.D., Consulting Psychologist, Berkeley, California, gives the details of a study he made comparing 22 rheumatoid arthritic patients with 22 controls. The controls were also patients at the same outpatient clinic but the "primary basis for choice of controls was the 'internal' syndrome as opposed to the 'external' as with rheumatoid arthritis."[41]

Subjects and controls were given various psychological tests, including the Minnesota Multiphasic Personality Inventory. In one summary Dr. Geist writes: "Because of the large number of scales used in analyzing the MMPI, a comprehensive personality assessment is indeed difficult. From the results of this test, the rheumatoid arthritics have weak egos, repress their hostility, express their psychopathology in physical symptoms and are likely to go into a psychotic state (under emotional stress). . . They seem to be more dependent, conscientious, compliant, masochistic, conservative, over concerned about their appearance, sensitive to others, busy and emotionally labile than the controls. . ."[42]

Under "Supplementary Findings," the author writes, "There were definite psychotic implications. . . The indications were that the psychotic manifestations were latent rather than overt and that these patients were being held together or prevented from going into a psychotic state by obsessive-compulsive defenses. . ."[43]

In a further summary, after utilization of Form B of the Holtzman Ink Blot Technique, with all forty-five blots being used, Dr. Geist writes: "The personality constellation of the rheumatoids as a group as delineated by the HIT. . . may be summarized as follows: They are relatively optimistic, with a capacity for warm interpersonal relationships and empathy with others. They tend to become disorganized in emotionally traumatic environmental situations (perhaps because of low ego strength). They conceive of their bodies as lacking a defensive wall and are very narcissistic. . . They do not see the world organized as a whole but tend to split it into pedantic parts indicating an obsessive-compulsive set of defenses."[44]

Dr. Geist writes further that he "feels that much of etiology is congenital both in the physical and psychological area. This predisposition based on heredity undergoes modifications due to experience. . ."[45] He further states that "the beneficial effects of a warm, dry climate are currently beyond dispute."[46] (In this connection, Dr. Arthur Freese, writing a decade later, reports in his book, *Help For Your Arthritis and Rheumatism,* that "there is *no* evidence that a warm, dry climate either cures or materially improves RA. In fact, Eskimos and Laplanders living in their intensely cold lands suffer less RA than those who live in warm dry climates!"[47])

Dr. Geist also writes that "some characteristics of rheumatoid arthritis are more responsive to placebos than are symptoms of most diseases. It has been estimated that up to 60 percent of rheumatoid arthritics are responsive to placebos while the effectiveness of placebos for all disorders is about 35 percent."[48] He suggests further that the chronic inhibited rage of these patients should be dealt with, but Dr. Freese, writing over a decade later in a book subtitled, "All the facts your doctor doesn't have time to tell you," in discussing the treatment of RA does not at any time mention any need to deal with inhibited rage.

In an experiment with three adult hemophiliacs with chronic arthritis, it was found that progressive muscle relaxation exercises and meditative breathing resulted in "significant maintained reductions in arthritic pain perception for all subjects."[49]

In another study it was found that with hospitalized mentally ill patients, the mortality rate from cancer was much higher than it is for the general public; also the morbidity from cardiovascular diseases is higher; prior to 1940 and the introduction of antibiotics the morbidity from infectious diseases was much higher; the risk of contracting Parkinson's disease is higher; *but the risk of being victimized by rheumatoid arthritis and allergic phenomena is lower.*[50]

What are we to make of all this?

Medical science, faced with a very old, very common ailment, seems uncommonly baffled by it. Rheumatoid Arthritis: Cause: unknown. Prognosis: uncertain. Treatment: guesswork. Cure: none. Many more women than men are afflicted and the medical profession seems not to have a clue as to why this is so.

One thing, however, they *do* know without question. RA is an *inflammatory* disease, characterized by *inflammation* of a joint or joints.

Inflammation: "1. The act or fact of inflaming; 2. the state of being inflamed. 3. *Pathol.* redness, swelling, pain, tenderness, heat, and disturbed function of an area of the body, esp. as a reaction of tissues to injurious agents."

Inflammation — the state of being inflamed.

SPECULATIVE COMMENT:

Arthritis is clearly an expression of the "state of being inflamed."

In addition to inflammation of the affected joints in RA, victims often suffer from the following:

Fever	A heating up of the blood, i.e., anger.
Weariness	It is tiring to inhibit rage.
Loss of appetite	Food = love; when we are caught in the grip of fury, we reject love.
Anemia	To be red-blooded is to be vigorous, virile, daring; to rigidly hold in anger is the opposite of daring.
Lungs may be affected	How we draw in life/breath is always affected by strong emotion.
with pleurisy developing	Pleurisy: inflammation of the pleura (a serous membrane investing each lung).
The skin, blood vessels, heart, eyes or muscles may be affected.	Of course.
Cold, tingling hands and feet	Fear, the flip side of anger.

Excessive sweating   Fear: cold sweat of fear.

RA may come on suddenly or slowly. Anger growing into barely controllable rage may come on suddenly or grow more slowly. Anger may flare up, be dealt with, and go away. Or it may flare up, be insufficiently dealt with and return. If it hardens into hate/resentment so that it becomes next to impossible to deal with, in time it "clings to its victim for longer and longer periods until it is present virtually all the time."

If RA is a defense against the release of anger, a defense against a psychotic breakdown, then if a psychotic breakdown has already occurred, there is little need any longer to defend against it; therefore, hospitalized, mentally ill patients would show a low incidence of it, as they do.

As the feminists have shown us all too clearly, boys and girls in our culture are subjected to entirely different social conditioning. Boys tend to be denied expression of many or all of the softer emotions but are allowed anger. Girls, on the other hand, are allowed all of the softer emotions but tend to be denied their anger. As a result many women grow up unable to face their angers, inhibiting their resentments and their rage to a much greater extent than men. If RA is an expression of repressed rage—if it strikes because someone has become so "burned up" that this inflammation is visited upon the joints—women would be much more likely to fall victim than men.

HYPOTHESIS:

Rheumatoid arthritis occurs when the mind becomes so full of repressed anger that it is no longer a proper fit for the spirit.

The ill fit is expressed in precisely those parts of the body where the lack of fit is occurring.

****

The spirit insists that anger is present and must be dealt with. The mind squirms away in fright, refusing to accept the message, and the physical body is left to manage on its own.

An infectious agent—anger—triggers the disease.

The body, absorbing the message that anger is there, mounts the only assault it can: against itself.

The joints are constructed to move but in a fluid, frictionless way.

RA victims have been shown to be overly conscientious people who inhibit their rage.

Resistance to movement produces heat, a proven fact of the physical world.

Resistance to movement produces inflammation, i.e., rheumatoid arthritis.

In RA the joints most often affected are in the fingers and toes.

We use our hands to reach out to others and often to work with.

I must do this but dammit I don't want to, or
I want to do this but dammit I can't.

We walk on our toes.

I must go this way but dammit I don't want to, or
I want to go this way but dammit I can't.

Other joints frequently affected:

Wrists:

I have to turn my hands to this but dammit I don't want to.

(Or the reverse.)

Knees:

I have to bow to others' will but dammit I don't want to.

Shoulders:

I have to shoulder this burden but dammit I don't want to.

Legs:

I have to move forward now but dammit I don't want to.

(Or the reverse.)

Elbows:

I don't dare stiff arm/straight arm—push away—anyone as much as I might want to.

Ankles:

I have to turn this way and that, even though dammit I don't want to.

Neck:

I'm sick of having to bend to others, dammit, dammit, dammit.

Jaw:

How I'd like to jawbone my way out of this, but I don't know how and don't dare even try.

An increasingly crippled body, to where the hands may become practically unusable and the feet badly distorted, to where tendons and muscles may no longer connect with bones so that movement becomes impossible, quite neatly solves all problems, but unfortunately, because this type of solution is not to the spirit's liking, the solution is experienced as an extremely painful one.

Let's take a brief look at another, even more common form of arthritis, "everybody's disease": osteoarthritis.

Osteoarthritis, more accurately termed "degenerative joint disease," is also known as the "wear and tear arthritis," or as "osteoarthrosis (indicating lack of inflammatory process)."[51] In osteoarthritis inflammation is rare, the primary problem being essentially degeneration.

This form of arthritis is also known as "senescent (elderly) arthritis," appropriately enough. "In the over-sixty age-group, an enormous 97 percent have severe enough OA for the disease to show up in x-ray films, and the remainder of the over-sixties may also have the disease but to an extent too slight to be seen in x-rays."[52]

Oddly enough, "it is possible to suffer from osteoarthritis and have perfectly normal x-rays."[53] The reverse is also true: degenerative changes indicative of OA can be revealed by x-ray in those who suffer no pain and have no overt symptoms of any kind.

Although primary osteoarthritis "appears in joints virtually simultaneously with the advancing years,"[54] nobody really knows the cause. Theories about causation include the possibility that hereditary factors are involved or metabolic abnormality, abnormalities of the endocrine glands, or inborn defects/mechanical imperfections in the joints.[55] "It is still a mystery, however, why some people get OA more readily than others or at a much earlier age or even why some are able to go through a long life with virtually no difficulty at all."[56]

Osteoarthritis is responsible for stiffness on arising, and the chief symptom of the disease is pain. With the onset of the disease, there is "a softening, pitting or fraying of the articular cartilage."[57] As the cartilage becomes increasingly damaged, movement causes stress, which leads to additional damage. "Eventually, as the OA progresses, whole sections of cartilage may be worn off, leaving the underlying bone exposed to the opposing bone in the joint so that bone grates on bone in movement. . . In very severe OA the entire inner structure of the joint may disintegrate and become distorted."[58]

The joints most commonly affected are the ones which bear the body's weight: the hips, the knees and the spine. "Also frequently involved are the joints at the tips of the fingers and at the base of the thumb and the big toe."[59] The jaw joints may also be affected, "the second or middle row of fingers joints and the joints atop the shoulder where the collarbone and shoulder blades meet."[60] Ankles are rarely affected "even though they too carry body weight. Rare, too, are problems in the elbows and shoulders, the wrists, and the joints at the base of the fingers."[61]

Fortunately, "OA is a relatively benign disorder. . . Pain, when present, is usually moderate and rarely becomes disabling. . . In short, osteoarthritis can be an annoying and minor rather than really troublesome disability for those fortunate enough to achieve a long life."[62]

QUESTION:

Is osteoarthritis properly considered a disease when x-rays reveal it even though there is no pain and no other symptom?

When pain is experienced but x-rays are perfectly normal, should this be considered osteoarthritis?

ANSWER:

Medical science seems to consider the disease present either on the evidence of x-rays or because pain is experienced, but is this a valid stand?

If "an enormous 97 percent have severe enough OA" for it to show up on x-ray, should this be considered a disease? The one thing that no human being seems able to achieve, over a long enough period, is to grow older without aging. If age brings certain changes in the joints, *but these changes are not accompanied by pain or any other symptom,* is this a disease? Is grey hair a disease? Are wrinkles? Is the more relaxed attitude and generally mellower outlook that many people experience with age a disease? Is having grown children and grandchildren and possibly even great grandchildren a disease? If aging itself is a disease, then of course any changes that commonly accompany aging would have to be considered diseases too.

SPECULATIVE COMMENT:

If the term "osteoarthritis" is used to describe the degenerative joint changes that commonly accompany aging, then the term should be restricted to this usage, should not be used to cover any other condition, and *should be classified as a disease only if all the changes that accompany aging are so classified.*

If pain accompanies the changes, or there are other symptoms, then the condition can be considered a diseased one, but *it should be designated by some term other than the term used to cover the normal changes brought about by aging.*

Inasmuch as "it is possible to suffer from osteoarthritis and have perfectly normal x-rays," it should be acknowledged that osteoarthritis *as a disease* is not dependent upon changes in the joints.

So what is OA as a disease dependent upon?

SPECULATIVE ANSWER:

Friction. The same kind of friction that in a more acute form results in RA.

If movement remains fluid and painless *in the mind*—in our thoughts and feelings—it will remain fluid and painless in the joints, regardless of any physical changes that aging brings.

****

Let's take a very quick look at the third most common form of arthritis, gout.

Whereas rheumatoid arthritis is predominantly a woman's disease, gout is a man's.

> "Hippocrates, the father of medicine, gave the first really clinical description of gout about 2,500 years ago. He outlined the patterns of the disease in three widely quoted aphorisms:
>
> "1. A eunuch does not take the gout.
>
> "2. A woman does not take the gout until her menses be stopped.
>
> "3. A young man does not take the gout until he indulges in coition."[63]

Right there we have all we need to know the underlying cause of this condition, also why this particular affliction, among all the infirmities man's flesh is heir to, has always seemed such a joke—to those who don't have it.

"For some reason," Dr. Arthur Freese writes, "gout has long provoked humor at the expense of its victims."[64] The reason seems simple enough: frightened by sex, man has always tended to snicker and smirk about it. The "cause of gout lies in the way the body deals with prolonged periods of excessive amounts of uric acid in the blood. . . Uric acid is produced in the course of the body's normal metabolic processes and is discharged chiefly in the urine."[65] Man can choose between urination and sex. If he chooses sex too obsessively, and with too much guilt, hello gout.

"For some unknown reason, the large joint of the big toe is the most common site of the first attack of gouty arthritis."[66]

This should not be an "unknown reason" for the reason seems clear enough. "Joint" is one of the slang terms for penis. Sex is a physical activity, largely performed by the physical part of the body (from the waist down), and the most physical part of this physical part of our bodies, the part that connects us directly with the physical earth, is the foot. For this reason as well as others, the foot has always been closely associated with the erotic. It is equipped with extremely sensitive nerve endings, and the involvement of the foot, in particular the toes, and most particularly the big toes, in sexual practices is far from unknown. "Toe sucking" is indulged in and advocated by the sexually sophisticated. ("Don't knock it until you've tried it," remarked one young woman with earnest belligerence when the practice was under discussion.) The big toe can be and occasionally is used as a substitute organ of copulation, primarily as a teaser.

If there is a build up of sexual guilt to where the pressure to punish becomes intolerable but the guilty one (self perceived guilty one) can't bear to risk his precious penis, what is the body to do? The penis can justifiably claim that it provides a life and death service, urination; that it is not in itself responsible for sexual excess, which begins in the head, reducing the penis to the role of helpless slave; and because the penis does have a redeeming physical value and while sexual excess may deserve punishment it nevertheless isn't a life and death matter, therefore it (the penis) should be excused from the block and a substitute put in its place.

What more natural substitute than the big toe?

The affected joint (the big toe in more than 75% of all cases) becomes "unbearably tender and painful with considerable swelling and redness. . ."[67] Typical engorgement, preparation for intercourse—except that instead of being accompanied by pleasurable feelings, the pain is excruciating, standard aversion therapy, a fitting punishment for real or imagined sexual sins.

Gout is a man's disease, but eunuchs don't get it, nor do young men until they have indulged in coition. Women are protected until their menses have stopped and in a procreative sense (and in this sense only) they are no longer the women they once were. Gout is commonly thought to be, and studies have shown it to be, an affliction of high living, a disease "ten times as frequent among famous Englishmen as among the British population as a whole,"[68] and naturally it's all a big joke to those unaffected by it.

SPECULATIVE SUMMARY:

All forms of arthritis are caused by a build up of repressed negatives to where the mind/body becomes an unfit habitation for the spirit.

HYPOTHESIS:

All disease—whether infectious or non infectious—is caused by a lack of fit between mind and spirit.

Every disease manifests in precisely that area of the body where the lack of a proper fit is occurring, which means that a knowledgeable reading of the body will tell us the *why* of any dis-ease.

Before we turn to a discussion as to how to properly read our bodies and in this way avoid disease, let's have a look, in the following chapter, at an affliction that is currently causing tremors in us all: cancer.

9

CANCER, THE MUTANT CELL

Under Too Rigid a Tyranny, Do Our Cells Rebel and Start The Revolution Without Us?

> It is not true that life is one
> damn thing after another—
> it's one damn thing over
> and over.
>
> Edna St. Vincent Millay

Cancer has become so prevalent that almost all of us have had some personal experience with it, if not in our own bodies then in the bodies of our loved ones or friends. Estimates are that over the years three out of four families will be stricken.

In the course of a year over 850,000 of us will be diagnosed as having cancer. It is the second leading cause of death following heart disease, and of every five deaths from all causes, one is from cancer. About 440,000 of us will die of our cancer this year.

In 1931 the medical profession confessed itself a failure where cancer was concerned. The best it could hope to do back then was cure one out of three cases.

Since then billions of dollars have been spent on cancer research, $5 billion by the United States Government and millions more by the American Cancer Society.

For our billions of dollars in taxes and donations, this is what we have achieved in fifty years:

The incidence of cancer is steadily increasing; therefore the expected mortality rate must be plotted on a rising curve. Whether figured in the total number of deaths each year or by the percentage of the total population, even more of us will die of cancer next year than this, even more of us the year after that, ad infinitum, unless and until we derail ourselves from the track we are on.

Fifty years of research, costing enormous amounts of money, hasn't saved us. Today we are more vulnerable than ever. What if anything will save us?

The term "cancer" is used to denote "any one of a group of over 100 related diseases characterized by the disorderly and uncontrolled multiplication of abnormal cells in the tissues of the body."[1] Primary types are *carcinoma,* which arises in the epithelial tissues (*epithelium: "Biol.* any tissue that covers a surface, lines a cavity, or the like and performs protective, secreting or other functions, as the epidermis."); *sarcoma,* which arises in the connective tissue; *leukemia,* cancer of the blood; and *lymphoma,* cancer of the lymph nodes.[2]

While the cause or causes of cancer is or are not yet known, numerous substances are believed to be or have been shown to be carcinogenic (*carcinogen: "Pathol.* a substance that tends to produce a cancer."). Primary among these substances is tobacco. ("Tobacco smoke, for example, is responsible for lung cancer and contributes to other cancers as well."[3]). Other carcinogens are said to abound in our environment: asbestos, vinyl chloride, tars, pitch, certain industrial oils, x-rays, radium, other radioactive materials, and certain food additives, to name the most widely recognized. Because of all these identified carcinogens, there is a growing tendency to see cancer as environmentally induced. However, viruses are also suspected in some if not all cancers. Age and sex play a part. Men are more frequently victims of some cancers and women of others, and for both sexes the older we get the more prone to cancer we are. Nutritional deficiencies may cause or be an important contributing factor in cancer of the liver and esophagus while over-exposure to sun increases the risk of skin cancer. There may also be an inbuilt susceptibility to certain cancers due to heredity.[4]

"The cancer cell is a profoundly abnormal cell. . . Its nucleus is larger and irregular."[5] It is not differentiated for function. "Cancer is also a disease of cell heredity. . . The deoxyribonucleic acid (DNA) of the nucleus, which stores and carries all the information that enables daughter cells to reproduce mother identically, is not the same in the cancer cell . . . some of the information represented in the DNA molecule is added and some is missing. In other words cancer cells are, strictly speaking, mutant cells."[6]

Normal cells divide until they come in contact with neighboring cells, then they stop. Cancer cells have lost this "contact inhibition."

"The destiny of the cancer cell is not to resemble its sisters or to obey the regulations of a superior order; the cancer cell is a wild and totally asocial individual, programmed to proliferate without restraint, to compete with its neighbors for food, and in the end, to destroy the organism at whose expense it survives."[7] Other cells may grow "in defiance of the organism's needs . . . , proliferate inappropriately and for no reason," producing a tumor. However, if the cells of the tumor "compress but do not invade" neighboring tissue they are classified as "benign, non cancerous."[8] If invasion of neighboring tissues and adjacent organs occurs, the growth is cancerous.

"Another essential characteristic of cancer tissue is that it does not stay in one place but reproduces elsewhere in the organism."[9] While some cancer cells remain where born, "others begin to travel. They infiltrate the blood vessels and circulate throughout the organism."[10] This process is known as metastasis ("*Pathol.* the transference of disease-producing organisms or of malignant or cancerous cells to other parts of the body by way of the blood vessels, lymphatics, or membranous surfaces.") This dissemination occurs very early, "during the first few days" of a tumor's development.[11]

"Invasion and metastasis are the two fundamental characteristics of malignancy."[12]

Cancer cells are believed to occur naturally in our bodies from birth on. Something goes wrong in cellular reproduction, believed due to the instability of DNA, "the chemical carrier of heredity," and a mutant cell with the capacity for uncontrolled growth is born.[13] Our immune system, alerted to destroy the malformed cell, usually succeeds with no disease resulting. But at other times, due to an as yet unexplained failure of the immune system, the mutant cells get a jump ahead of their attackers, aren't destroyed, and "then we have what is called *cancer,* the clinical term for a process always taking place, *but now out of control.*"[14]

That bacteria might be involved in the causation of cancer has been suggested for decades, only to be rejected by the entrenched medical establishment. Early pioneers—Dr. Michael J. Scott of Butte, Montana, Dr. Thomas J. Glover of New York, and Dr. Glover's laboratory assistant, Tom Deaken—experimented in the early twenties with an anti-cancer serum with marked success, so much so that the *Anaconda* (Montana) *Standard* declared in headlines on Sunday morning, June 15, 1924: DR. SCOTT TELLS OF CURE FOR DREAD CANCER PLAGUE.[15] In 1940 Dr. Glover, in collaboration with a surgeon, Dr. John E. White of Malone, New York, published a report detailing their successful treatment of 328 cancer patients from the early 1920s to 1939. They published their report privately after it was decreed by the National Institutes of Health that publication by the Public Health Service would wait upon a repetition of their work. "It is obvious, even at this distance, that every attempt was made to suppress this historical confirmation of the bacterial relationship to cancer that Dr. Michael J. Scott had announced in detail in 1924."[16]

Another researcher, Dr. W. M. Crofton, Lecturer in Special Pathology, University College, Dublin, Ireland, published a book in which he claimed that "a microbe was always associated with cancer." Dr. Crofton "made vaccines from the bacterial forms, which he used with surprising success. . . Crofton's work was bitterly attacked, and he found it impossible to publish articles, though others confirmed his technique and results."[17]

Experiments conducted early in this century by a cancer researcher, Peyton Rous, suggested a possible bacterial or viral causation for cancer. Working with chickens, Rous was able to transmit sarcoma "by inoculating with dead cells or, in the absence of cells, by a tumour extract that had passed through a Berkefeld filter. Subsequently Rous found two other varieties of sarcoma in hens which could be transmitted by filtered tumour juice. In each of the three instances the special characters of the original growth were repeated in the experimental tumours. He and his collaborators brought forward strong evidence that the filterable agent was a living but extremely small microbe." However, that cancer might be caused by a virus was an extremely unpopular notion with those in power, even back in 1910, so Dr. Rous was derided and ridiculed and his experimental results pushed out of view and quickly forgotten.

Bacterial involvement in cancer has apparently been confirmed by more recent research. Dr. James Hiller, an expert in electronic microscopy, states categorically (as quoted in a paper by Dr. Virginia Wheeler and her colleagues), that "a microorganism is present and can be demonstrated with the proper technique in every type of human tumor and in animal tumors we have studied. . ."[19]

Dr. Wheeler, a "prominent researcher who has dominated the field of cancer microbiology,"[20] has made two startling and possibly profoundly significant findings.

"In 1965, a paper by Dr. Wheeler and Dr. (Eleanor) Alexander-Jackson reports that the chemical actinomycin was found in the urine and in cultures from cancer patients. The significance is that the most likely source of this toxic chemical compound were bacteria from the *Actinomycetales* group."[21] Drs. Wheeler and Alexander-Jackson "proposed the name of *Progenitor cryptocides* (ancestral hidden killer) for this bacterium."[22] This was a bacterium that could be classified as a virus as its life cycle includes a state so small that it cannot be detected with standard microscope techniques but must be viewed under an electron microscope.[23] In Europe, other researchers, working with cancer patients, isolated and observed a microbe with identical characteristics.[24]

If these findings are borne out by future investigators, then an anti-cancer serum—such as the one used successfully by the early cancer pioneers back in the 1920s—could very probably be developed which would wipe out cancer as effectively as the cowpox vaccine did away with smallpox.

Dr. Wheeler's second major discovery has to do with *chorionic gonadotropin* or *choriogonadotropin*—CG. CG is a substance synthesized and secreted by the trophoblast ("*Embryol.* the layer of extraembryonic ectoderm that chiefly nourishes the embryo or develops into fetal membranes with nutritive functions.")

When the sperm enters the egg within the female body and the cells begin to divide, the developing embryo has chromosomes from the father, which means it is subject to attack as a foreign body. One of the functions of CG is to ward off this attack. This is accomplished through a "high density negative charge. Agents of the immune system also carry a negative charge, but of lesser magnitude."[25] Like charges repel each other, so the CG effectively protects the "foreign intruder," the new life, from being assaulted and routed by the body's immune system.[26]

In 1976, this substance (CG) was also found in human spermatozoa.[27]

"Since 1970, different investigators have also found the presence of a CG-like substance in the serum of a great number of patients with cancer, suggesting that the cancer cells, like the trophoblasts and sperm, are also producers of CG or a CG-like substance."[28] Further study confirmed this: cancer cells do indeed synthesize the CG-like compound.

In 1974, Dr. Virginia Livingston-Wheeler announced "a major medical research breakthrough: her own discovery and confirmation that the microbe of cancer also synthesizes a CG-like material much like the material synthesized by cancer cells."[29]

Current research indicates that "CG or the CG-like protein is found *only* in cancer cells, spermatozoa, trophoblasts of pregnancy, and cancer-related bacteria."[30]

This finding suggests 1) that cancer is not a group of 100 to 300 related diseases as is insisted upon by conventional medical theory but is one disease with many related forms; and 2) that there is an excellent chance that an effective anti-cancer serum can be developed.[31]

If cancer *is* one disease, and cancer cells ward off assault by the body's immune system with CG or a CG-like protein carrying a heavy negative charge, then a serum capable of chemically neutralizing the CG-protein would enable the body's immune system to do its job: get at the mutant cells to attack and destroy them. The rising incidence of cancer would be reversed and cancer could become as obsolete as the deadly plagues of the middle ages.

An effective anti-cancer serum was first developed and used in 1924. Why are we still struggling with near epidemic levels of the disease almost sixty years later?

Regarding an anti-cancer serum, Dr. Robert E. Netterberg and Robert T. Taylor write in their fascinating book, *The Cancer Conspiracy,* "This immensely promising line of research (anti-cancer serum) . . . is the direct result of over eighty years of research into cancer related bacteria—research that has been swept under the rug, denied and mocked while thousands of victims have died unnecessarily in the terrible pain of cancer. And it is staggering to think of what may lie ahead when this nation finally realizes what has happened since 1924."[32]

We voluntarily donate money for cancer research and are taxed to support additional research, with a rising curve of new cases and a nearly stagnant cure rate in repayment.

"All of the experimental work to develop the bacterial CG-cancer cell relationship and a universal vaccine against cancer is being sponsored strictly through private funding! Not one dime of money . . . is coming from the powerfully financed cancer establishment!"[33]

Meanwhile, the conventional medical treatments for cancer are surgery, radiation and chemotherapy, all aimed at cutting out or killing off the malignant cells without destroying too much surrounding tissue. A somewhat more controversial approach is immunotherapy in which bacterial agents are used to stimulate greater immunity in the host body. For the past few years we have been hearing a great deal about a protein produced by the body, *interferon* ("*Biochm.* a protein substance produced by virus-invaded cells that prevents a reproduction of the virus.") Interferon was being publicized as a possibly powerful new weapon in the fight against cancer after early studies found that it tended to inhibit the growth of tumors and to stimulate the immune system to work against them. However, just recently it was reported that a two-year study conducted by Dr. Shelby Berger of the National Cancer Institutes and three co-workers found that interferon, instead of inhibiting the spread of cancer, may actually help the cancer cells to spread into normal tissue.

As to possible preventive measures against cancer, we are warned on all sides against the use of tobacco. Nutritionists also warn against the additives in prepared foods, the hazards of our impure drinking water, and hypothesize that cancer may be a deficiency disease:

"Scientists can tell . . . whether cells are mature or not. And they have found that as a vitamin B_{12} deficiency grows worse in pernicious anemia, the red blood cells become more and more immature in appearance and character. In cancer, too, as the disease progresses, the cancerous cells become more and more immature, seeming to indicate that less and less of some important 'maturant' substance is present as the disease goes on."[34]

On Sunday, December 13, 1981, the Baltimore Sun ran a story by Isaac Rehert under the heading, A DOCTOR'S ENCOUNTER WITH CANCER, with the subheading, "Macrobiotic Diet Cured Him, Hospital Head Believes." Dr. Anthony Sattilaro, a fifty-year-old, self styled "classical conservative," chief executive of the Methodist Hospital in Philadelphia, learned in his annual physical checkup in 1978 that "cancer was pervasive throughout his body: in the skull, the right shoulder, probably the sternum, the left sixth rib and the genital region.

"As a doctor, he recognized a death sentence when he saw it. Early in June, he admitted himself to Methodist Hospital and surgeons began removing cancerous tissue."

Three operations were performed in three weeks. "Everything . . . removed was found to be riddled with cancer.

"Doctors put him on large dosages of estrogen, which made him itchy, caused him to vomit a great deal and to gain weight. He was in near constant pain, so daily he took a concoction consisting of morphine, cocaine and compazine."

His father was also battling cancer, and on August 7th that year lost the battle and died of it. "It was on the trip back from the funeral in New Jersey that Sattilaro began his own return from death. Driving a rented car on the New Jersey Turnpike, depressed and weary from pain, he picked up two youthful hitchhikers.

"Their names were Sean McLean and Bill Bochbracher, and when Sattilaro told them he had just buried his father and that he himself was dying of cancer, McLean said to him, 'You don't have to die; cancer isn't all that hard to cure,' and proceeded to tell him about the macrobiotic diet." About a week after this encounter, Dr. Sattilaro received in the mail from McLean a book called *A Macrobiotic Approach to Cancer.* Figuring he had nothing to lose, Sattilaro decided to give it a try.

"The basic macrobiotic diet consists of 50% cooked whole grains (brown rice, wheat barley, millet and others), about 25% locally grown vegetables, 15% beans and sea vegetables, and the remainder fish, soup, condiments, fruits, seeds and nuts." After embarking on the diet, the doctor began to feel better. "Once, on a vacation where he couldn't get fish, he violated his regimen by taking a few bites of chicken instead. He promptly became ill again. From that time on, he never veered from his diet." He began using chopsticks to eat with, gave up wearing synthetic materials in his clothing, against advice stopped taking the estrogens that had been prescribed, continued to feel fine—"in fact, he believed he was getting stronger every day."

In September of 1979, "he returned to oncology for a rerun of the tests he had been given the year before. The tests could find no trace of cancer in his body."

"Today, he said, he is working more than ever, although with more ease and less effort than he ever used to use."

A close friend and co-worker, Dr. John Giacobbo, vice president of the Methodist Hospital, Philadelphia, calls himself a "supportive skeptic. . . If I believed in miracles, I'd say what happened to him was a miracle. . . But I don't believe in miracles. . . I don't know how to explain what happened. He definitely had cancer—I've seen the evidence—and now he's definitely clean."

A nutritional approach that worked—for one man at least.

Other nonconventional treatments include Krebiozen and Laetrile, both of which have been pronounced worthless by the American medical establishment but continue to have their passionate and dedicated adherents.

"Dr. Stevan Durovic, a former assistant professor at the University of Belgrade . . . came to the United States after several years of experimentation in South America, with a drug which he had discovered and which he believed to be a cure for cancer."[35] He was put in touch with "Dr. Andrew C. Ivy, a medical researcher of unassailable reputation and gigantic stature in his profession. Dr. Ivy was, at the time, vice president of the University of Illinois, and head of its huge Medical School. His efforts in cancer research had led to his appointment as Executive Director of the National Advisory Cancer Council and he was also a director of the American Cancer Society. An important, a respected, an honored man."[36]

Dr. Ivy, interested in Dr. Durovic's substance, Krebiozen, decided to test it scientifically in order to get "an accurate picture of the drug's potential. From the very beginning the results were astounding in their positiveness. Though Krebiozen was used only on persons who had been diagnosed as hopeless and close to death, its remarkable characteristics showed themselves almost at once. There was a lessening or complete disappearance of pain, and in many cases tumors were dissolved and replaced with healthy tissue. Physicians who were trying it all over the country reported like results. It began to look as though science had finally come up with a weapon against cancer that had a chance to win the fight."[37]

So what was the fate of this promising new cancer treatment?

"In rapid succession Dr. Ivy was suspended from membership in the Chicago Medical Society, removed from the vice presidency of the University of Illinois, and had his resignation accepted by both national cancer societies noted above."[38] His work with Krebiozen "was assailed as inaccurate and unscientific,"[39] and it was even suggested that he had become senile. A "Status Report on Krebiozen" was published in the Journal of American Medicine, a report shot through with fraud, according to later investigation by Herbert Bailey, who wrote two books on the subject: *K-Krebiozen—Key to Cancer?* and *A Matter of Life or Death, The Incredible Story of Krebiozen.* The doctor who wrote the damaging article, according to Mr. Bailey, "made so many omissions of factual and favorable results and so altered the findings that were at his disposal that even the most naive researcher would have discredited his conclusions at once as throughly unscientific and worthless."[40] Nevertheless the status report—valid or fraudulent, scientific or not—carried the day and, medically speaking, Krebiozen was dead. Those who were making millions from the suffering of cancer patients could continue to do so.

Treatment with Laetrile combined with diet, as practiced in the clinic and hospital of Dr. Ernest Contreras in Tijuana, Mexico, is another approach which the American medical establishment has pronounced worthless. Anyone journeying there, however, will run across numerous people, returning for their semi annual or annual checkups and drug supplies, who claim they have been helped or cured by this regimen. After receiving death sentences from American doctors, who pronounced them incurable, they headed to Mexico seeking hope and help, and, according to their enthusiastic testimony, received both.

Another controversial approach is the Simonton method, developed by Dr. O. Carl Simonton of Fort Worth, Texas. Dr. Simonton believes that psychological forces play an important, possibly crucial role in the development of cancer and that these same forces can be rallied and used to overcome it. To mobilize these psychological healing forces, he has developed a special mental technique called "imaging." The cancer victim is taught "healing visualization," wherein he visualizes an inner battlefield upon which healthy cells drive out or destroy malignant ones.

In the view of Dr. Simonton and his partner, Stephanie Matthews-Simonton, there is a cancer-prone personality with identifiable characteristics. These characteristics include the inclination to cling to resentments, be unforgiving, indulge in self pity, find it difficult to develop close, enduring relationships, and suffer from low self esteem. For the Simontons, no one "*gets* cancer."[41] There is a wish to withdraw from life and in pursuit of this the choice is made to develop cancer. "Malignancy is thus despair that has been experienced biologically, despair at the level of the cell. Cancer is a state of despair made manifest and real."[42]

The Simontons run a Cancer Counseling and Research Center in Fort Worth where patients enroll for a ten day program. In addition, professionals across the country who work with cancer patients have been trained in the Simonton method. The Center is thriving, the claim is that about 100 patients "have outlived their life expectancies," and some who have graduated from the treatment program speak glowingly of it and insist they've been helped.

Predictably, the response of conventional medicine has ranged from labeling the method unproven to condemning it as unethical and dangerous. "The game here, so Simonton's appalled critics assert, should be called, 'Blame the patient.'"[43] Dr. Simonton himself admits that his treatment approach can engender guilt, "stripping down defenses and hurting people more than you're helping."[44] A controlled study of 160 women admitted to King's College Hospital, London, carried out by Steven Greer and Tina Morris, failed to confirm earlier reports of associations between cancer and stress, in particular the loss of a loved one, habitual denial of stress and depression. This study did not substantiate Dr. Simonton's contention that cancer is a physical manifestation of a state of despair. Also, since cancer is known to develop in many different species, dogs, reptiles and fish among others, and even in plants, it seems implausible that emotional factors would predominate over biological ones.[45]

To attempt to summarize the above is to find oneself faced with a mountain of conflicting opinion and data.

Cancer is a term used to cover 100 to 300 related diseases, according to conventional medical theory, but some leading researchers in the field are convinced that the demonstrated presence of the protein substance CG, or a CG-like protein substance, found on *all* cancer cells is proof that cancer is one disease only which manifests in different forms.

Not only is it not incontrovertibly known what cancer is, it is also not known what causes it. Recent thought has swung toward the environment as the primary causative factor, yet other researchers, those tracking down the presence of CG, offer evidence that it is bacterially induced. Dr. Lucien Israel, a leading French authority, writes that, "One might regard cancer not as a disease that strikes by accident but as an almost inevitable consequence of aging."[46] Our immune systems wear out with age and cancer results. Nutritionists suspect that cancer may be a nutritional deficiency disease, while Dr. O. Carl Simonton among others attributes it to psychological causation.

The American medical establishment claims that its treatment methods, surgery, radiation and chemotherapy, are the only proven ones, yet we have made little advance beyond the 1931 cure rate of one in three while the incidence of the disease continues to rise and more of us die of cancer each year.

A serum treatment used successfully in the 1920s failed to stir up sufficient interest for the medical establishment to do anything further with it and, falling into disuse, it has been for the most part forgotten.

Proponents of Krebiozen claim that this drug, under testing, proved to be remarkably effective, but this treatment too, derided by the medical establishment, soon fell by the wayside.

Meanwhile, we keep getting new cancers in record numbers and the majority of those stricken will be killed off by the over production of these malformed, mutant cells who have lost all "contact inhibition" and rebelliously go their own way at our expense. Production of cancer cells is a normal bodily function, but somewhere along the line too many of us are losing our ability to keep them under control.

Is there any way we can regain greater control and keep what should be a momentary brush-fire flare-up of rebellion from growing into an all out, deadly revolution?

Is there any way at all we can protect ourselves from this near epidemic onslaught?

SPECULATIVE COMMENT:

We have already hypothesized that all diseases—infectious or non infectious—have only one cause: a lack of fit between mind and spirit. Whether cancer is a non infectious disease, as is currently believed by the powers that be in our medical establishment, or will in time prove to be a low grade infectious disease, as some researchers now claim, is immaterial to our argument. When a misfit occurs, the spirit can summon whatever means work best to bring the lack of fit to the mind's attention. We don't have to know the particular means used in any given disease.

Researchers have found that a protein substance CG (chorionic gonadotropin or choriogonadotropin) or a CG-like protein substance is found on the surface of spermatozoa, fetal cells, *and on cancer cells.*

The sperm are protected, the embryo is protected, to ensure the safety of new life. Our bodies continually manufacture new cells as old ones die, *but only cancer cells are afforded the unique protection of being coated by CG.* For some reason, these malformed, mutant, uncivilized cells are put under the same umbrella of chemical/electrical protection provided for the sperm and the fetus. Cancer, no matter how destructive it may prove to be, *is treated within our bodies as a welcome invasion of new life.*

HYPOTHESIS:

From birth on, we produce cancer cells (an established fact), but cancer *as a disease* develops only when the tyranny under which the cells live has grown so unbearable that rebellion comes as a welcome relief, a welcome manifestation of new life in protest against the old.

****

To see how this works, let's take a close look at the leading cause of cancer deaths among women, breast cancer.

Lucien Israel, M.D., in his book *Conquering Cancer,* writes, "In males, cancer occurs more often before twenty and after sixty; in females it is more frequent between those two ages, because of the incidence of cancer of the breast and uterus."[47]

This statement not only informs us of an interesting situation, it also leaves curiously unexplained what it is talking about.

Because of the incidence of breast and uterine cancer, women don't follow the same age patterns that men do in falling victim to cancer, but instead develop cancer more frequently between the ages of twenty and sixty.

Why *because?*

Women have breasts and uteruses *before* the age of twenty and *after* the age of sixty. If one might regard cancer, as Dr. Israel writes, "not as a disease that strikes by accident but as an almost inevitable consequence of aging," why doesn't this correlation with age hold true for women when it comes to their most frequent cause of cancer deaths, cancer of the breast?

Instead of correlating with age, breast cancer strikes frequently between the ages of twenty and sixty, *correlating with fertility rather than with age,* a rising incidence occuring a few years after fertility begins and dying away a dozen or so years after fertility ends.

Studies have shown that teenagers rarely develop breast cancer.

Breast cancer strikes childless women more frequently than women who have borne children.

To turn this second, well established statistic around: bearing children affords at least some protection against breast cancer.

Why does childbearing offer this increased protection?

Cancer cells are mutant cells, rebellious cells. Rebellions are staged when a government becomes unbearably repressive, when the governed either have their potential thwarted and suffer from non use, or are wrongly used and suffer from misuse.

HYPOTHESIS:

Cancer of the breast is a rebellion against non use or misuse of the breast.

****

The female breast serves two closely related functions: it is a sexually attractive feature; it is the organ which produces and secretes milk for the nourishment of the young.

Biologically speaking, the former function is subservient to the latter. That is, biologically speaking the purpose of being sexually attractive is to attract a mate in order to propagate.

To phrase it another way, the breast has only one primary function, which is to produce and secrete milk for the nourishment of the young.

The female breast, uniquely among all bodily organs,is a source of food, of nourishment as love: maternal love. If the breast is not allowed to fulfill this function, if maternal love is rigidly repressed, a rebellion of mutant cells is finally staged as a measure of last resort, the only hope of throwing off the tyranny of repressive fear.

Teenage girls rarely develop breast cancer not because they don't have breasts—they do—but because the crying need to express maternal love is not yet upon them; the future holds too much promise. Rigid repression of maternal love has not yet had time to install itself.

Between twenty and sixty, women deal with the needs of their bodies, their minds and their hearts, either by having children and loving them or not doing so. By the age of sixty this challenge has for the most part been dealt with, one way or the other, so the incidence of breast cancer falls off, not because women no longer have breasts, for in the absence of surgical removal they *do* still have them, but because the primary breast function is no longer at issue.

Childless women repress maternal love more frequently than do women who have borne children and therefore they are more vulnerable to cancer of the breast.

Though women who have borne children are less vulnerable, such women, in fear, can still rigidly repress love toward a given child or children—can fail to face and deal with their love or lack of it for a given child or children—and in this way leave themselves open to the uninhibited march of rebellious cells.

For example, a woman may give birth to a child and lose this adored child in infancy, a devastating blow to her heart. When she gives birth a second time, she may dam the free flow of love to the second child in fear of suffering another devastating loss. To live ruled by fear, allowing fear to block the free flow of maternal love, leaves a woman open to a rebellious uprising: breast cancer.

For a variety of reasons, women who have borne children may find it too painful to love them freely, and breast cancer may result.

The above is non use of the breast.

Misuse of the breast occurs as follows:

Though there has been a great deal of ferment lately and some change, nevertheless for centuries women have been culturally conditioned to play a passive, subservient, sex-object role vis-a-vis men. This is in sharp contrast to the maternal role, which ideally is an active, protective role, a role similar to that which the male in our society is conditioned to play vis-a-vis the female.

In many ways the maternal role calls for traits and behavior which our society, in all its wisdom, has labeled "masculine."

If a woman, regardless of how many children she has given birth to and mothered, is afraid of this active, protective, "masculine" role, if she uses her breasts as a sexually attractive inducement but in fear never moves from this passive, subservient, sex-object role into a more active, dynamic, "masculine" role, she is in effect allowing the secondary role of the breast to predominate over the primary role, and this constitutes misuse.

Fear of this more active, dynamic, "masculine" role—a fearful clinging to a more passive role, leading to an over dependence on men and on male approval—is usually caused by, or translates into, a fear of lesbian feeling.

To reword and make more specific our hypothesis:

HYPOTHESIS:

Breast cancer is caused by a severe repression of maternal feelings, or a severe repression of lesbian feelings.

Cancer of the breast proclaims, in effect: "You failed to use your breast properly, therefore, we, your rebellious cancer cells, will now take over and run things ourselves."

These wild, rebellious cells have a profound, unassailable logic on their side. If we fail to use or misuse anything, how long can we legitimately expect to remain in control? Cancer signals a loss of control.

"Claus Bahnson, a psychoanalytic researcher, has noted 'conformism' and 'religiosity' as prominent personality features of women who develop breast cancer."[48]

In a culture where women are taught to be passive, receptive, to be sexual objects, non aggressive, a cultural standard which finds its roots in and is reinforced by the Christian Church, the more a woman conforms and the more seriously she takes her religion, the more difficult it will be for her to fulfill the maternal/masculine function of her breasts, and the more susceptible she will be to a cancerous rebellion.

****

Whereas cancer of the breast is the leading cause of cancer deaths in women, the leading cause of cancer deaths in men is lung cancer.

Cigarettes have been indicted as a leading causative factor in lung cancer; in fact, it is common to run across statements in print that cigarette smoking "causes" lung cancer, which is a very broad if not careless use of the word "causes." While a link between smoking and lung cancer has unquestionably been established—"The risk of lung cancer of a non smoker is not zero, but it is thirty times less than the risk for a man who for fifteen of twenty years has been smoking two packs of cigarettes a day."[49]—this link is very clearly no direct cause and effect relationship. We all know people who have smoked for fifty or sixty years and who die peacefully in their beds at eighty-nine, and of others who have never smoked yet who die of lung cancer at sixty-two.

In the flood of statistics we have been given establishing the link between cigarettes and lung cancer, no one has ever told us precisely or even approximately how many cigarettes must be smoked before lung cancer results. We are told that to smoke at all increases the risk not only of lung cancer but of many other cancers, yet it is easy to prove observationally that it is possible for some people to smoke for decades without ill effect, and even the statistics show that if smokers give up the habit, their risk of lung cancer is in time no greater than the risk run by those who have never smoked. A link between the inhalation of cigarette smoke and lung cancer clearly exists but *a linkage does not prove a causal relationship.*

Phenomena can occur concurrently without there being a causal relationship. When we have a cold, we can be bothered by sore, itchy eyes and a runny nose. The soreness in our eyes does not cause the nasal discharge or vice versa. Both are symptoms of the same underlying cause, the cold we have.

Cigarette smoking and lung cancer, like itchy eyes and runny noses, may be linked only in that both are symptoms of a deeper underlying problem.

SPECULATIVE COMMENT:

Cigarettes are a phallic-shaped object. Primitive peoples often eat organs of choice from the bodies of those they have slain as a way to incorporate into themselves the admirable qualities of their fallen enemies. In the Catholic Mass, the flesh of Christ is ingested in symbolic form. When we smoke a cigarette we are symbolically taking into ourselves the phallus.

But this isn't all we are doing. As we smoke and take in the phallus, we are at the same time destroying it through fire.

If there is a deep dissatisfaction over gender identity, if fury or fear is at work, we can express it by becoming heavy cigarette smokers and/or developing a growth or growths in the lungs, ie, lung cancer. Neither one causes the other; both are symptoms of the same underlying dissatisfaction, fury or fear.

Gender identiy is a fundamental identity. The moment a baby is born, as it is slapped or otherwise prodded to take the first breath, the sex is noted. The first thing we ask about a new arrival, once we hear that both mother and child are alive and well, is: boy or girl? Therefore problems with gender identity, a fundamental identity, express themselves in the organ that processes the most fundamental requirement in life, breath, the breath that is life.

HYPOTHESIS:

Lung cancer is an expression of anger/fear over gender identity problems.

To phrase it another way: Lung cancer is the body's response to a craving for increased masculinity, or a less vulnerable masculinity, a craving caused by anger/fear.

****

Statistics show that women fall victim to lung cancer far less frequently than men, even when they are heavy smokers.

The dissatisfaction and anger that some women feel at not being men, at not having the rights and prerogatives that adhere to the male in our society, can be fierce enough to make heavy, chain smokers out of them. But if this heavy smoking sufficiently expresses the fury—the two dozen times a day such women ingest the penis, take on penile attributes, while at the same time destroying an endless chain of penises, burning them up, blowing them out contemptuously in smoke—then the heavy smoking won't lead to any malignant growth in the lungs.

If smoking does not successfully express the anger, with its reverse face of fear, then the remaining fear and anger must be dealt with, quite possibly through a cancerous growth in the lungs.

Smoking expresses the same anger/fear for the male but with much greater risk, for this reason. Even as the heavy male smoker ingests a great number of penises to boost his uncertain gender identity, at the same time he is destroying, burning up, the very increased penis-power he feels most in need of. This means that smoking can exacerbate the underlying fear quite as much as it eases it.

For both sexes, smoking is an angry protest against gender expectations. The male smoker is saying, in effect: if only all this hogwash about the "masculine" would go up in smoke, then I could stop my incessant worry that I'm not "masculine" enough. Both sexes are protesting the myth of masculine "superman" invincibility, but for the male, as a rule, it is a far more intense and personal protest. While both sexes are to some extent victimized by the myth, the male is subjected to greater stress.

That our cultural stereotypes place men under greater stress is statistically borne out in several ways. Men die at younger ages, more often falling victim to stress-related illnesses. More men commit suicide, become criminals or alcoholics. More men are transvestites and transsexuals. It may not be easy to be a woman but evidence from every direction proves it is even more of a strain to be a man.

To research the question as to whether stress and cancer are linked, Robert J. Keehn of the Medical Follow-up Agency of the National Academy of Sciences National Research Council studied the health records of former World War II and Korean war prisoners. He failed to uncover any heightened cancer death rate. If stress and cancer are linked, if stress contributes to cancer, then wouldn't these former prisoners of war, subjected as they were to long periods of abnormal stress, have fallen victim to the disease at a higher rate than the general population?[50] The study showed that they didn't so any stress-cancer link was considered disproved.

All fine and good, but there are surely varying kinds of stress.

One stress that would not be increased under prisoner of war conditions is the stress caused by uncertainty about one's proper gender behavior.

For all the deprivation and fear suffered by prisoners of war, one uncertainty they are not up against is what constitutes, in a prison camp, proper "masculine" behavior. Instead there would be reasonably clear, straightforward standards as to what it means, under those conditions, to "be a man" or to "act like a man."

But what does it mean to "be a man" or to "act like a man" in our "free" society? Even with the advancing tide of women's liberation, men are still expected to work and provide, or help provide, support for their wives and children, to be "good providers." So how much should they lie, cheat, and trample on the rights of others to achieve this? How much should they "go along" to "get along"?

In his delightfully entertaining book, *Donahue, My Own Story,* Phil Donahue writes: "For me the biggest shock of growing up, learning what it means to be an adult breadwinning male in America, was learning about ass kissing."[51]

If it's necessary to kiss ass to get along, how much ass kissing is enough? How much is too much? At what point should a man stop sacrificing his self respect as he struggles to live up to his expected role of good provider? *Help, someone, I don't know the rules; where do I go to learn them?*

Even in our day, after two decades of effort on the part of some to achieve "liberation" for everyone, men are still expected to wear pants in the family, at least one pair of them even if his wife also wears them. Few types in our society are as laughed at, sneered at, looked down upon, as the henpecked husband, the male who is perceived as being at the mercy of a balls-busting wife. So how should a man deal with the problem if he finds himself married to a woman who doesn't show him sufficient respect, who back talks him or lets the kids get away with a lot of gruff? If he can't get his outrage across any other way, should he fall back on his greater physical strength and give her a punch in the mouth to make his point? If that doesn't do it, should he beat her up? Will that assure him a proper role in the family, make a *man* out of him? Or does such bullying behavior simply prove how weak he is, how little the man he ought to be? *Help, someone, I'm not sure of the rules; where do I go to learn them?*

Surely few things can be more stressful than to be forced to play an extremely serious game, a game in which one's self image and reputation will be forged, without being given an adequate understanding of the rules of the game. Yet all of us, men in particular, live under a heavy barrage of ambivalent messages, to the point that it is no wonder a certain percentage of men loathe the very idea of "masculinity" and live with the churning fury that if only they could get enough of the stuff—prove to be enough of a "man"—they could then turn the tables on the damn masculinity concept and destroy it, as they would happily give their right arms to do. Smoking expresses this anger/fear, but unfortunately often costs a man his lungs and his life instead of an arm.

Not all symptoms of any given affliction are always present. It is possible to get a severe cold without suffering from a sore throat or developing a cough. People can struggle with a great deal of unresolved anger/fear about gender identity and gender expectations and yet for one reason or another never take up smoking. The only symptom expressing their underlying problem may be a malignant growth in the lungs.

"The rate of lung cancer among younger subjects apparently remains relatively stable," Dr. Lucien Israel writes in his book *Conquering Cancer.* "If this tendency is confirmed, what does it mean?"[52]

It means that the world is loosening up a little, stereotypes are being broken, the old days when men were expected to be MEN and women were assumed to like it is being challenged and in some quarters defeated. It means that both sexes are now more willing to allow men to be human, as opposed to being hard, erect, silent, stoic, money-making penises, while women struggle to be accepted as fully human too.

When women feel a deep down satisfaction with being women, no longer are frightened or angered by it, accept it fully, they will either not smoke at all or will quit the habit. When men find it possible to relax into their own masculine gender without fury or fear, fully accepting themselves, they will either not smoke at all or will quit the habit. And when these conditions are met, when both men and women can relax and enjoy, lung cancer will be no more.

Without anger/fear operating within us, there would be no cancer or any other disease.

Let's turn to a discussion as to how to achieve and/or maintain physical health in this frantic, fearful world.

10

BODY-SPEAK

Tune In, Turn On, and Don't Drop Out—
Until You're Ready

> To become a thoroughly good man is the best prescription for keeping a sound mind and a sound body.
>
> Francis Bowen

Inscribed in letters of gold on the temple at Delphi, the site of the most famous oracle of the ancient world, was the admonition: *Know Thyself.*

"Knowledge is power," said Lord Bacon, and certainly the more thoroughly we understand and know anything, the greater is our power and our control.

SPECULATIVE COMMENT:

If we would keep our thoughts and our feelings in proper fit with our spirit, we need to dive deep within to get to know what is boiling away down there. *Know Thyself.* While this is easy to say, it is not the easiest job in the world. It may be the hardest.

The Christian Bible makes the claim that perfect love casts out fear. This ties in with the widely held belief that hate is the opposite of love. If we accept the notion that hate is a particularly painful form of anger, anger solidified, and that fear is another aspect of anger, we arrive at the conclusion that love and fear are opposites, as light and darkness are opposites: where one exists, the other is banished. Love casts out fear; fear casts out love.

Let's assume for the sake of discussion that we all live somewhere along the scale between perfect love and complete and abject fear. If we lived totally at the love end, one hundred percent loving without fear, we would be living saints, too light, too good for this world, while if we lived totally at the fear end, one hundred percent afraid without a glimmer of light or love, we wouldn't last long enough to worry about it—the weight of the fear would crush us. So somewhere between the two extremes of total love and total fear we struggle along, striving and crying, expressing, repressing, and generally making a botch of things.

Definitions:

Love is a sense of unity/oneness, while a sense of separation—isolation—alienation equals non love, or anger/fear.

A simplified hypothesis regarding the causation of disease:

HYPOTHESIS

A movement toward love fits the spirit; a movement away from love, toward increased anger/fear, ill befits the spirit and causes strain, a strain that results in physical disease.

To rephrase:

All disease is caused by the unwillingness—or, more precisely, the inability—to love.

To put it even more simply:

Love is the basis of good physical, mental, emotional health.

Or

Love is good health.

Fear is the absence of good health.

Pain is the language the indwelling spirit uses when speaking to us, to get the message across to us.

****

QUESTION:

But if the indwelling spirit is so gung ho on love as to push us insistently toward it, why doesn't the spirit show a little love itself and speak directly to us in our conscious thoughts instead of indirectly, sneakily, though bodily pain?

SPECULATIVE ANSWER:

The spirit tries; in fact, is in the process of trying to get the message through to the mind and heart when the body intercepts the message and pitches in to try to help.

Consider this situation:

A painter enters a room intent on painting the air. He sets up his spray gun, fills it with yellow paint, aims it at the air around him, pulls the trigger and sprays. The air becomes cloudy and yellowish, but every time the painter turns off the spray, the air clears rather quickly again. *It does not stay painted.* But the paint from the gun splatters onto every solid object around and soon the floor, ceiling, and walls are bright yellow, as are all pieces of furniture in the room. The painter has not achieved what he set out to do: he has not successfully painted the air. Instead he has accomplished what he did *not* set out to do: he has painted the room.

In like fashion, consider this scenario for the birth of a cold:

The spirit, suffering with the negative emotions piling up, poisoned by them, begins broadcasting a message to the conscious mind that these negatives are present and must be dealt with. Electrical impulses zoom across the brain, but the airy mind, which would rather not know about such things, darts here and there, shrugging off the droplets of paint. The yellow awareness falls onto nerve fibers, these absorb the message and send it out: "Anger is here, express it." The throat begins to warm up with guilt, the nose begins to tickle as it gathers moisture to sneeze. Anger is loose in the body and will be expressed through rejection, on the mental/emotional level on which anger lives. Cold germs are everywhere, can be summoned almost entirely at will, put to work, and lo and behold we come down with a cold—*because the body hears and absorbs the message that the frantic, darting mind has been too frightened and squeamish to deal with itself.*

In support of this view of the birth of a cold, consider two rather curious findings of the Salisbury Common Cold Research Unit and other researchers.

One of the "age-old myths about colds" that Dr. Andrewes, director of the Salisbury Common Cold Research Unit, set out to test was the almost universally held belief that chilling is a factor in the production of colds. Almost all of us have grown up with mothers who admonished us to wear our rubbers when it rained so we wouldn't catch cold. We are urged to come in out of the rain before we catch our death from a cold. To become wet and chilled is seen as a sure way to invite a cold into our systems, as all but putting out the welcome sign. Such a universally held belief would seem all but certainly rooted in experience, rooted in fact.

But Dr. Andrewes, in his zeal to test all aspects of cold production, tested this out too, and found that, under controlled conditions, it didn't hold up. Six volunteers "took hot baths and then stood in a cool hallway for half an hour, wearing wet swimsuits. Then they dressed but wore wet socks." None of the six caught cold.[1] "In another experiment, when the volunteers were thoroughly chilled by walking in the rain and returned to cold rooms, there was no indication that chilling had made them sensitive to the (cold) virus; and all succeeding experiments on chilling continued to be negative."[2]

What are we to make of this? Under controlled experiments, chilling failed to produce colds. Another cherished myth bites the dust. Chilling does *not* "cause" colds.

But how could we have been so wrong for so long? Everyone *knows* that getting caught in a rainstorm, getting soaked, being forced to remain in wet clothes, becoming thoroughly chilled, leads to sore throats, sniffles and coughs. We have experienced it ourselves too often. How can science "disprove" what we *know* to be true?

Even Dr. Andrewes had trouble accepting these findings. In the face of "proven" negative results, he continued to believe that weather, or some upset having to do with seasonal changes, played a part in the production of colds. "If some sort of stress is necessary to lower resistance, then it is not surprising that these experiments failed. The Salisbury volunteers were probably not sufficiently stressed. As a matter of fact, they found the experiments quite amusing."[3]

Stress.

What kind of stress?

Consider this scenario:

A man leaves his office building after having worked, and worried, hard all day to find to his distress that the heavens have opened up and are pouring down. His most presentable business suit is in danger of getting soaked, his old shoes are springing cracks, and, dammit, with the new muffler he *had* to put on the car leaving him flatter than usual, and with the cost of repairing the neighbor's picture window that Jimmy so carelessly broke still hanging over his head, and why couldn't Molly have kept a closer eye on the kid, he should never have been playing hardball in the street anyway, and when at least eight kids were playing, shouldn't all the parents ante up to pay for the damn window, after all any one of the boys—or either of the two girls playing—could have hit the damn ball that broke the window, and my agent told me to get liability insurance, I should have listened to him, and dammit my right sock is already soaked through, I've got to get new shoes—not after next payday, *today!*—these goddamn old ones are a disgrace, if the boss were to see them—oh, Christ. Why does the parking lot have to be half a block away, I'm soaked already, all I need is to come down with a cold, why didn't Molly, who keeps the radio on all day, or the damn TV, she must have heard the weather report, why didn't she phone to warn me it was going to rain? When the washer breaks down she can call me fast enough, but when she might be of some help, I could have slipped out a few minutes early, missed this goddamn downpour, dammit, dammit, dammit. Sneeze, sneeze, sneeze.

If when we are out in the cold, we *feel* "out in the cold"—abused, rejected, unloved—it's extremely easy to tap into the accumulated anger which, accompanied by guilt, is expressed by a cold. Self pity moves into resentment/rage and the sneezes begin.

In contrast, when Dr. Andrewes used his volunteers to test whether chilling led to colds, the experience was viewed as an amusing lark. The volunteers were there to be of use to science, to be of service to their fellowman, all in the spirit of helpfulness and fun. So if the staff wanted them to take hot baths, then stand in cool hallways in wet swimsuits, so what? What was the worst that could happen? A cold. Well, that's what the volunteers were there for, to catch colds. So why not? How easy to stand out in the cold without in the least *feeling* "out in the cold." As long as no anger was felt, being chilled failed to lower resistance. There was no resentment around to toss the welcome mat out for the cold to step in.

The second rather curious fact discovered by cold researchers was that bodily defenses against cold germs were most easily breached when the germs entered *through the eye.* We can breathe in cold germs with relative impunity, eat them, swallow them, even have them injected into us and our chances of coming down with a cold remain fairly slight. But let the germs enter through the eye—presto, a cold.

SPECULATIVE COMMENT:

The eye is located on the mental level of the body and, connecting directly with the brain, is the facial feature that is mental on the mental level.

It is through the eye that we most often, most easily catch our colds.

We *think* ourselves into anger, accompanied by guilt, and cold germs become our servants, allowing us a safe, non threatening way to express the rage that we feel.

****

QUESTION:

If we'd prefer not to become ill, even with a cold, how do we go about this?

ANSWER:

We listen.

If we are free of all pain, the spirit and mind are in easy fit and all is well. If they fall out of fit, which causes dis-ease, the spirit will let us know, one way or another.

If all other ways fail—as they usually do, the mind being the frantic, flighty, experienced-at-dodging coward that it is—the body will pick up the message and pass it along in the form of pain.

When all else fails, pain becomes the language with which the indwelling spirit manages to speak to us.

If we would be well, we must listen.

Listen to our bodies, to the pain they feel.

We don't have to fear, as we first begin to listen, that we'll have any trouble hearing. If the pain our bodies feel is a real pain, it won't come through overly soft or muted so that we have to strain to hear. Rather, the "sound" of pain ordinarily comes through loud and clear, so much so that far from having to quiet down to hear it, we most often wish we could somehow shut it off. But we all know it's dangerous to drug ourselves past a certain point to shut off pain, and at other times it's useless—the pain persists regardless of how many pills we take. So here's an alternative suggestion:

Learn the language of the body and listen to it, if you'd prefer not to be ill.

HYPOTHESIS:

To remain healthy, all we need do is have our conscious minds deal with and discharge negative emotions instead of shrugging them off onto the body.

This way *can* be chosen and used, if we decide to make the effort. However, if we've been living for years with an avoidance-prone mind which is an experienced dodger-of-all issues, a cowardly mind which for decades has forced the body to carry all burdens, it will take time, patience, and above all intelligent application to effect a successful switch-over from body to mind.

Where to start?

For purposes of illustration, let's start slowly, easily, with the common cold.

The air that we breathe in—the breath of life—after being separated in our nostrils is reunited in our throats. The throat is a symbol of unity, of the oneness of life which is love. Any violation of love produces the burning redness of shame in the throat.

A sore throat equals guilt.

Guilt over what?

Oftentimes guilt over growing anger, rising hostility.

Sneezes, coughs, a runny nose often express that hostility.

If you want to avoid the misery of a cold, take note of the very first fleeting symptom you feel. A tickling sensation in your throat? A slight discomfort when you swallow? A series of sneezes? A slight stuffiness in your head? Whatever it is, *pay attention.*

As you *hear* the symptom, ask yourself why it's there. If the initial discomfort is in your throat, it's because you feel guilty. Quite possibly a build-up of anger is precipitating the guilt. Why are you feeling hostile and toward whom? Drag it up into the conscious mind. Face it there, discharge it there, and you won't have to express it through bodily discomfort, through a burning throat, a heavy, throbbing head, a persistently runny nose, a congested chest, and/or all of the other distressful cold symptoms.

If you're not accustomed to facing your negative feelings, if your pattern has been to avoid them, to shrug them off onto your body to express them for you, don't expect to develop instant rapport between your conscious mind and your unwelcome negative feelings. If you've been repressing angry/fearful thoughts for years, they are by now so thoroughly beaten down, so afraid of the light, they are going to take a bit of coaxing before you can get them to creep out front and center. But it *can* be done—and once you do it, once your conscious mind begins to grapple with your angers, relieving your body of the burden, you *can* expect very quick to instant alleviation of your bodily symptoms.

Our major problem with facing our hostilities is that they are so often sparked by and directed toward those we most cherish, those whose love we most crave. We fear our loved ones won't continue to love us if we admit how angry we feel toward them. This is a pattern begun in childhood when we fear Mommie will disapprove and may even—terror of terrors!—desert us if we dare to express how angry we feel toward her. In a panic we press down the anger, out of mind, out of heart, and develop a headache or tummyache instead, and Mommie rushes to hold and comfort us, crooning her love and concern, which quickly teaches us the rewards of illness. No one blames us for being sick, while being angry is something else. Very early we develop a pattern of not dealing with our negative feelings and converting them into physical dis-ease.

As adults, when we come down with colds we are doing the same thing, pushing negatives out of mind. Rather than deal with our angry feelings, it seems far easier and safer to come down with a cold, for which no one will blame us. Our dearest beloved will take tender care of us, reassuring us that we are loved as we sneeze, spit and cough out our hostilities. But the indwelling spirit sees this as a cowardly way out, and any time we are ruled by fear we suffer for it. For this reason, coming down with a cold is rarely a truly pleasant experience.

QUESTION:

Isn't the above playing the same game that critics of Dr. Carl Simonton accuse him of playing, a rotten, "Blame the Patient" game?

ANSWER:

Of course. It is placing upon the victim of any illness full blame—full responsibility—for any dis-ease he suffers. It is taking this position: if the mind/heart/spirit occupying a body is not fully responsible for its care and upkeep, who is?

****

When we give thought to the advantages of suffering with a cold over the uncertainty and panic many people feel about facing their own angers, it is easy to see why this particular illness is perennially popular. Current prognostications that the common cold will never be conquered may prove to be accurate. However, if you wish to banish colds from your own life, there is really very little to it. *Learn the language of your body and listen to it.* What you deal with in your conscious mind you won't have to deal with in your throat, nose, head and chest.

A helpful hint: Anger/fear are two faces of the same negative emotion. We use anger to keep from having to face our fear. If you find your body giving you the signal that a cold is approaching if you don't discharge the load of hostilities building up, skip right over asking yourself why you're feeling angry and toward whom, and get right down to bedrock: *what are you afraid of?* If you are building up anger, you're afraid of something. Face the fear and the anger will dissipate.

When it comes to the common cold, it is easy to speak glibly of dealing with the negatives involved in the conscious mind instead of in the body. The destructive power being dealt with is minor either way. As miserable as we feel with a cold, within a few days we will no doubt recover, usually no worse for wear, and the negatives involved, if dealt with consciously instead of in the form of a cold, are not going to break the mind. We are not working with potentials powerful enough for real destruction.

When we turn to a severe affliction such as rheumatoid arthritis, however, we are taking a step toward much greater potential destruction. Studies have shown that arthritis may be used, by some victims, as a defense against psychotic breakdown: the negatives involved seem so ferocious that a crippling of the body seems preferable to a possible mental/emotional crackup. Most of us, if forced to choose between these two alternatives, *would* rather suffer from inflamed joints, no matter how severely this hampered us, than to have ourselves institutionalized as psychotics.

Scientists working with fissionable material in the development of the A Bomb were forced to exert the utmost care. Converting mass into energy is not a risk-free procedure. THIS IS ALSO TRUE WITHIN THE BODY. Once a conversion of energy into mass has been made—once the mind has shrugged off negatives onto the body to handle—*the reconversion process is a high risk venture.*

Another way to phrase this is that it is easier to stay healthy to begin with than to regain good physical health once it has been mindlessly thrown away.

SPECULATIVE COMMENT:

Despite the risk factor, reconversion *is* possible. If undertaken patiently and with sufficient care, it can be accomplished without so overloading the conscious mind that it blows or breaks under the strain.

Keeping the above in mind, let's take time for a brief review of rheumatoid arthritis:

RA can develop slowly over many months or years or can on occasion strike suddenly, with severe pain and disability. "Most often, however, RA follows a deceptively gentle course. The victim has symptoms for a few days, but they disappear. They reappear with more severity later, then fade away once more. When they return again, it is with a new fury. Days or weeks or months may pass between these flare-ups and remissions, but gradually the disease clings to its victim for longer and longer periods until it is present virtually all the time."[5]

If we would prefer not to be crippled by RA, we need to pay attention at the first sign of pain. In most cases, RA gives the victim repeated warnings before it settles down into the body for the long haul. If we don't want to invite RA in for an extended stay, WE MUST LISTEN TO OUR BODIES AND PAY ATTENTION.

At the first sign of an inflamed joint, we should call to mind the meaning of the inflammation. *Resistance to movement causes heat.* This is an established physical fact. If we are hurting in our fingers, if our finger joints are becoming inflamed, we are resisting what our fingers are up to. What are we forcing ourselves to do that we don't want to do? Or, conversely, what do our fingers itch to do that we are afraid to let them do? If the pain is in our feet, where are we headed that we don't want to go? Or, conversely, where do we want to go that we are too frightened to allow ourselves to go? Wherever the joint inflammation occurs, that's the precise area in which resistance is causing heat. Our frantic, escapist minds have pushed the battle off onto the appropriate joint. Drag the conflict back into the mind, resolve it there, and the joint inflammation will fade away.

Pain is a warning. The warning contains a message. Once the message has been delivered, the pain goes away.

Acute pain has long been recognized as the body's warning system. *Chronic pain serves exactly the same function. The pain has become chronic simply because the message contained in the pain has yet to be successfully delivered.* Chronic pain is a telephone set to ring forever until someone answers it, or a tireless visitor at the door who will continue to knock until the door is finally opened. We can stop chronic pain this way: by answering the phone or opening the door. We must listen to our bodies and learn to read their messages if we would have chronic pain go away.

To interject here a pessimistic note: In physics there is a law formulated by Robert Hooke known as the Law of Elasticity which states that a force working on a stationary mass produces distortion in the mass. Remove the force and the mass will snap back into its original shape. If the mass does not resume its original shape, *the force has distorted the mass beyond its elastic limit.*

No substance is totally elastic.

Stretch a rubber band and let go. It will snap back into its original size—*almost but not fully.* Stretch it hard enough, long enough, and it won't snap back at all. The force applied to it will have exceeded its elastic limit.

This law of elasticity unquestionably applies to our internal world quite as much as it applies to the external world. Force applied too long and too hard to our working parts will distort them past the point where a release of pressure will allow them to snap back. The damage done may prove to be irreparable.

This does not mean that we cannot improve our health by releasing the distorting pressure. We can. But the degree of improvement may depend to an unhappy extent upon the amount of irreparable damage already done.

What of those life-threatening illnesses, such as hypertension and many forms of cancer, which do not play fair with open declarations of war but instead sneak up on us below our pain threshold and wrestle us onto the mat without prior warning of any kind? How can we deal with them?

With difficulty, admittedly.

With heart disease and cancer we are dealing with energy to mass conversions which are occurring on a very deep and lethal level. To ensure our own safety, our mental, emotional and physical health, we are almost forced to live up to Francis Bowen's prescription, quoted at the start of this chapter, to become a "thoroughly good" person. But even if we are willing to strive toward this, how can we know for certain what being a "thoroughly good" person entails? How can we know, before it's too late, whether or not we've succeeded? When we are seemingly at the mercy of repressed angers/fears within ourselves, how can we come up with an accurate appraisal of how good or not good we are?

We can pay attention to the admonition inscribed in letters of gold at the Temple of Delphi, also quoted at the start of this chapter: *Know Thyself.* With adequate self knowledge, we can catch ourselves operating under the pressures of rage or fear and in this way protect ourselves even against insidious subterranean exchanges that all too often have fatal results.

Take cancer of the breast, for example, caused by non use or misuse of the breast.

Women may not know when they are repressing maternal love, for that is the nature of repressions: we are not consciously aware of what we repress. But any woman who takes the time to check herself out will know how she feels in her heart about having children, or how dependent or independent she feels with regard to men.

If a woman tells herself glibly that kids are all right but that she herself does not want any as they wouldn't fit into her lifestyle, but half a day later, feeling instant strain, quickly averts her eyes from the sight of a newborn babe, or is too busy to attend the birthday party for a niece or nephew, and finds it simply *too* boring to listen to the endless chatter of women engrossed with their children, she should check herself out more closely: is she honestly facing what is in her heart? If tears flood her eyes at the oddest times, if she feels a sudden inner tension at the sound of a young child laughing or crying, or if the thought of one particular child always sends her mind running, scrambling away, she had better stop running long enough to face her tension and fear. The cells of her breasts are receiving the impact of these energy impulses which she is frantically shoving out of her mind, and any force acting on a stationary mass produces distortion in that mass.

With a minimum of introspective effort, most women know whether they feel basically independent of men or overly dependent upon them, though quite possibly this is not perceived as a yardstick upon which to measure fear of latent lesbian feelings.

Years ago on a TV game show, "Tattletales," three married women were asked if there was anything more important in life to them than their husbands' approval. Two said no, there was not; the third said her husband's love was more important to her than his approval. *Not one of them said that self approval, self love, was more important.*

Women have been socialized to express feelings that they often do not really have, and the three wives referred to above may have simply been responding by rote to what has been culturally fed into them. However, women do unquestionably exist who cannot feel good about themselves, who feel rotten and depressed, isolated and out in the cold, lost in the darkness, if they are not basking daily in the sunshine of male approval. This degree of dependency is an affront to the dignity and power of the female breast and constitutes abuse. Any woman that dependent on men should look within herself to find out why, to face the underlying angers and fears that keep her in such an enslaved state.

Any woman who trembles at thought of losing her lover, who will compromise herself to cling to some man, any woman who makes the highly charged claim that she would rather see her beloved dead than in the arms of another woman, is sitting on a powder keg of unfaced fears and should attempt to deal with those fears before they harden into rebellion in one or both of her breasts.

Lung cancer and the male:

Men who feel oppressed by masculine stereotypes, weighted down with the heavy expectations placed on them, will know, if they look inside even briefly, that these feelings are there. If their passionate involvement in sports has a slightly phony feel, if they are made extremely uncomfortable by the sight or thought of two men embracing, if they glance around and grow livid at the suspicion that other men are possibly showing interest in them, if they find themselves frantically blinking back tears that spring irrationally to their eyes, they can at least suspect that they are struggling against stereotypes which they find too rigid. If they can then take this a step further to the realization that the rigidity exists *within* themselves, and is symptomatic of underlying fear, this will give them a handle to work on. Work on themselves enough and this will lessen or end the energies being shot down to the lung cells. The message "loosen up, soften up," once absorbed by the mind and acted upon, will no longer force such a change in the tissues of the lungs.

In general if we would stay fit and pain-free, we must practice a detached scrutiny of ourselves, check on our dreams, our Freudian slips, and above all on those irrational tensions which grip us, letting us know that fear is at work. If we do this, making sure that our angers and fears are dealt with, not by a repression of them, sweeping them under the rug, but through a conscious awareness and expresison of them, in a large measure we will protect ourselves from a destructive conversion of energy into mass, will keep the forces within us from exerting a distorting effect on our bodies, and in this way will enjoy the manifold delights of good health.

SUMMATION:

The indwelling spirit may seem elusive, hard to keep up with, difficult to communicate with, but if we listen to our bodies and pay attention to ourselves enough to spot tension and the telltale signs of fear, we can learn how to cleanse ourselves so that we remain or become a proper fit for the spirit.

****

While the spirit may be somewhat difficult to communicate with, it is extremely easy to understand, for it operates under only one rule.

That rule is love.

A movement within us toward love fits the spirit. A movement away from love, toward increased anger/fear, ill befits the spirit and causes strain.

If we can follow this one movement/rule in our thoughts, in our feelings and in our behavior, we can assure a proper mind to spirit fit and enjoy good health.

If we were capable of driving out all anger/fear, if we were able to let ourselves be flooded at all times with a sense of unity and love, we would never experience even a moment's worth of uninvited—unwelcome—pain.

11

VERIFICATION

To Test or Not To Test, Here Are Some Questions

> Ultimately, our troubles are due to dogma and deduction; we find no new truth because we take some venerable but questionable proposition as the indubitable starting point, and never think of putting this assumption itself to a test of observation or experiment.
>
> Will Durant

Throughout this book speculative comment has led to the formulation of hypotheses. For these hypotheses to be elevated to theory, and then possibly to the lofty heights of accepted law, they must be corroborated—or disproved—through observation and testing.

The thrust of the preceding chapters has been toward a view of man as totally responsible for himself. This view runs counter to the prevailing winds of our times. As this century wears away, people are held responsible for fewer and fewer aspects of their lives. Criminals are not *truly* responsible for the crimes they commit. Either they suffer from "diminished capacity," often due to the ingestion of mind-altering drugs, or they are the helpless victims of impoverished backgrounds. Those convicted of crimes will repeatedly voice their regrets not by saying, "I am sorry I committed this act," but, "I feel terrible that this happened," or, "I'm sorry that this happened to me." If we are alcoholics, we are the victims of a disease. If we are compulsive over-eaters, we are of course victims. No matter what goes wrong, it is beyond our control. No one is at the helm; we view ourselves as buffeted mercilessly about by the vile, malicious winds of chance, victims of our genes, of germs, and of our external environment.

This book suggests, in chapter after chapter, that such a view is, and ought to be seen as, untenable.

No doubt it is comforting, in a way, to view oneself as a victim. It's a neat, easy way to dispense with all guilt. I'm not responsible for this mess I find myself in: my genes—or germs—or unfavorable external circumstances did this to me. No matter that this puts us in, and keeps us in, an enslaved condition.

Only as we accept full responsibility for every aspect of our lives do we have any hopes of being free.

No one who can be victimized by forces external to himself can be considered free.

Only one fully in control, beyond the reach of anything external to himself, can be considered free.

If we would breathe in the heady, intoxicating air of freedom, we must first accept the notion that we create our own lives and are fully responsible for all the circumstances of those lives.

That we do in fact create our own lives and are fully responsible has been the underlying theme of this book.

Any number of studies could be done to prove or disprove this contention.

In the opening chapter I attempted to show that our bodies tell us that our lives depend upon oneness/unity.

In later research I found, to my delight, that the underlying oneness/unity of life has already been demonstrated, and is accepted as proved, by some scientists working with subatomic particles.

In his fascinating book, *The Tao of Physics,* Dr. Fritjof Capra writes:

> "In ordinary life, we are not aware of this unity of all things, but divide the world into separate objects and events. This division is, of course, useful and necessary to cope with our everyday environment, but it is not a fundamental feature of reality. It is an abstraction devised by our discriminating and categorizing intellect . . .
>
> "The basic oneness of the universe is . . . one of the most important revelations of modern physics. It becomes apparent at the atomic level and manifests itself more and more as one penetrates deeper into matter, down into the realm of subatomic particles. The unity of all things and events will be a recurring theme throughout. . . As we study the various models of subatomic physics we shall see that they express again and again, in different ways, the same insight—that the constituents of matter and the basic phenomena involving them are all interconnected, interrelated and interdependent; that they cannot be understood as isolated entities, but only as integrated parts of the whole."[1]

Dr. Capra quotes from an article by David Bohm and B. Hiley:

> "One is led to a new notion of unbroken wholeness. . . We have reversed the usual classical notion that the independent 'elementary parts' of the world are the fundamental reality. . . Rather, we say that inseparable quantum interconnectedness of the whole universe is the fundamental reality, and that relatively independently behaving parts are merely particular and contingent forms within this whole."[2]

SPECULATIVE COMMENT:

Man is one of the "relatively independently behaving parts" mentioned by Bohm and Hiley in the above quote, but how often do we see ourselves as this, as "merely particular and contingent forms within this whole"? How often do we rule our lives with full awareness that "inseparable . . . interconnectedness of the whole universe is the fundamental reality"?

Dr. Capra suggests that not being aware of the unity of all things, instead dividing the world into separate objects and events, has been "useful and necessary to cope with our everyday environment." But, lacking awareness of this underlying unity of all things, how well has man coped? Look at our history. Recurrent wars, plagues, crimes, exploitation—this is *coping?* One wonders what the history of the world would have been had we not made this artificial division of the world "into separate objects and events" and in consequence not been able to cope.

The fundamental reality of life—some physicists tell us—is interconnectedness, interrelatedness, interdependence.

If life depends upon oneness/unity—as our bodies indicate it does—then any assault on oneness/unity becomes an assault against life.

If we see our neighbor as fundamentally separate from ourselves, and this distorted view of him justifies our right to strike him, rob him, or otherwise harm him, we are striking at our own life roots as well as at his.

If we see another nation as fundamentally distinct and separate from our own, and the populace of that nation as more wrongheaded than we are or as somehow not quite as human, and this distorted perception justifies waging war, in warring against such a nation and its populace we are warring against ourselves, for the fundamental reality of life is that all of life is one. We cannot slaughter our "enemies" without slaughtering an aspect of ourselves. We cannot wrong another without wronging ourselves.

****

In Chapter II there is a discussion of the observable connection between urination and ejaculation.

Whether this is simply a plumbing convenience without any underlying significance, or instead exists due to a meaningful correlation between the two activities, is a matter easily studied.

HYPOTHESIS:

The frequency of urination correlates inversely to the degree of sex drive and consequent sexual activity, *in those individuals subject to sexual pressures, for whom sexual activity remains a viable alternative.*

There are those few among us who are natural celibates, from birth on destined to be "old maids" or chaste, virginal bachelors, for whom sexual activity has absolutely no appeal or who find it so terrifying they forswear it, and among these few, whether or not they take up a religious vocation which requires celibacy, the correlation would not apply.

If you have a water pipe leading into a faucet with two outlets, where the water is channeled first into one outlet and then the other, with a built in barrier to prevent usage of both outlets at the same time, the frequency of the use of one outlet will correlate inversely with the frequency with which the second outlet is used. So it is with sex and urination. But if one of the outlets has been permanently plugged, never to be used, the open channel can be used at whatever frequency seems natural and convenient. So it is with celibates.

To test the above hypothesis, all that is needed is a questionnaire designed to elicit information about the two functions under consideration: the degree of sex drive, the frequency of urination.

PREDICTED OUTCOME:

Those who perceive themselves as having a high sex drive and whose sexual activities bear this out will also have a talent for holding their urine, will not possess over active kidneys.

Those who perceive themselves as having a low to middling sex drive, and whose sexual activities bear this out, will have healthy, but bothersome, over active kidneys. One or two drinks of any liquid and they will be glancing nervously around for the nearest restroom while their more sexually driven friends have no such concern.

Frequency of urination varies inversely with sexual drive/ sexual control.

To state it another way: those who can easily control their sex drive have a difficult time controlling the annoying frequency with which they have to urinate.

Those who can easily control the frequency with which they urinate have a more difficult time controlling themselves sexually.

If a study of these two functions finds that the hypothesized relationship exists, the implications are obvious, not only with regard to kidney disease but also with regard to assisting the sexually driven to gain control.

SPECULATIVE SUMMARY:

Kidney infections are expressions of sexual identity problems.

Kidney failure is the failure to resolve deep seated, bothersome sexual problems.

Resolution of the underlying sexual problems could aid in the recovery of normal kidney function.

In reverse fashion, with the sexually driven a new, two pronged approach could be tried. In concert with the long, circuitous, often ineffective route through the head, there could be an attempt to stimulate the laxy, inactive kidneys through physical means. As the kidneys become more active, this should lessen the lack of control over sexual drive.

****

Speculations about color tie in with speculations about how little or how much we are at the mercy of our genes.

If we form our own bodies to suit our own needs, as hypothesized, then one of the things we decide for ourselves is our coloring.

Studies could easily be done correlating eye color with personality type or choice of occupation.

Let's consider two common eye colors: blue and green.

SPECULATIVE COMMENT:

All colors have numerous associations and connotations.

As the color of the sky, blue becomes the color of peace and heaven, of heavenly peace. But as few of us reach this heaven while stuck here on earth, blue becomes the color of the wild, unreachable beyond, the color of defeat and depression.

To be blue eyed is, in slang terms, to be "innocent, gullible, idealistic, unworldly," while to be blue-nosed is to be strongly puritannical. Another use of the color mocks this innocence and Puritanism: blue comic material/blue jokes are lewd, lascivious, erotic, risque, vulgar, suggesting the obscene.[3]

To be blue eyed is to be at peace, or to thirst for peace, which makes one a visionary or an escapist. How would this translate into occupational choices?

Green is the color of hope, of spring, of renewal, but though the earth renews itself with such bursting vigor each year, man often finds himself unable to do so and jealousy and envy set in.

To be green eyed is to have an inner liveliness, restlessness, envy, jealousy, to feel a need to spring up and exert control. What occupational choices would this type of temperament find most congenial?

PREDICTION:

If studies were done correlating eye color with occupation, a higher percentage of blue eyed people than is found in the general population would be found in these occupations:

Artists of all types—writers, painters, sculptors, actors—*except* for musicians; journalists; physicists; physicians except for surgeons and psychiatrists.

A higher percentage of green eyed people than is found in the general population would be found in these occupations:

Composers, orchestra leaders; surgeons, psychiatrists; dentists; politicians and those who deal in illicit drugs.

****

As we build our own bodies to suit our own needs, we do this not only through coloring but also through form. Studies could easily be done to verify or disprove this.

Let's start at the bottom, with our feet.

Our feet are part of our physical selves; they are our direct connection to the earth.

If a study were made of foot size in relation to personality type, the following would be found:

PREDICTION:

People who are strongly materialistic, who care about material possessions and about their position on the economic/status ladder, who revel in sex and are driven by sexual needs—a personality type typified by politicians and drug dealers—will be found to have large feet in relation to their height and general body build.

People who are non materialistic, basically indifferent to their physical surroundings, who can say in truth about themselves that they could live happily in a log cabin in the woods or a tarpaper shack on the desert, people who have little real interest in sex and are in no way driven by their sexual appetites, will have small, dainty feet in relation to their height and general body build.

****

With our hands we reach out to others and to the world, to grab or to give. It takes a larger hand to grab successfully with than to give with. Materialists will have larger hands in relation to general body size, non materialists smaller hands in relation to general body size.

****

The waist divides our physical selves from our emotional selves. Those for whom love is founded primarily in sexual passion will have negligible waistlines; those for whom the deepest, most profound love is divorced from sexual passion—those for whom parental or brotherly or religious love is experienced as the deepest of all loves—will have far more pronounced waistlines, ie, will have much smaller waists.

****

The female breast symbolizes maternal love. A positive relationship exists between breast size and the depth and force of unexpressed maternal love. Women who have little or no maternal feeling to express will have little with which to fill a bikini top. Women who give birth and happily express their maternal feelings will find their breasts decreasing in size, and tending to sag; there is little inner excitement and anticipation left within to keep them high and firm. A compulsion to mother which has been blocked in expression will create overly large breasts, the larger the breast the greater the blockage in expression.

As women pass through menopause and begin edging into old age, another factor sets in. The breast symbolizes new life, and few women who outlive their fertility go springing forth to seek out new life in other forms. It's easier to continue along in the rut one has already carved out, no matter how dull and boring or even miserable that rut is. To jump out of the familiar and go off to challenge the unknown takes more energy, more emotional get-up-and-go than most middle-aged women have. Fat accumulates in areas where there is impaired function. Therefore older breasts are heavier breasts, for most women.

For those older women who either climb out of their ruts or get forcibly thrown out, who then create exciting new lives for themselves, this breast heaviness won't develop. The new life charging through them will keep their breasts young.

****

The penis is the primary male sexual weapon. In the eyes of children, sexual intercourse is seen as an attack on the female by the male. Our slang shows clearly that this conception of sex as an attack remains with us long after we have left childhood behind.

Penile size will vary as to how worried or indifferent a man feels about being armed with a dangerous weapon. For those who forswear use of the weapon against the woman they most love and fear—Mama—and through her by extension to all women, it does not matter how large their weapon is, for in the depths of their being they have unloaded it. This means that natural born celibates and homosexuals can grow large penises. Men who are incurably chauvinistic, who in their deepest hearts feel superior to women and have not the least worry about assaulting them, will also grow large penises for themselves.

If a man is not a natural born celibate, or a homosexual, or an incurable chauvinist, the more sensitively he worries about attacking women with this fearsome weapon he possesses, the smaller will be his penis. If he can't quite give up his right to attack, but is fearful of hurting anyone, to keep the size of his weapon small becomes the best compromise he knows how to make.

****

We look upon burdens as something we shoulder. If life feels very burdensome and we worry about our ability to carry our share of the load, we grow broad shoulders. If life seems more of a lark, an adventure, and we have no fear of holding up our end of things, we have no need of broad shoulders and grow narrower ones.

Studies could easily be done to verify or disprove any of the correlations between bodily features and personality type outlined above.

****

With regard to diseases, let's start with the common cold.

We can experience what is known as the common cold in a variety of ways.

Those who feel fortunate, who see life as generally good, will feel guilty as they get ready to come down with a cold and will suffer from severe sore throats.

Those who feel unfortunate, who see life as having treated them shabbily, will have only minor soreness in the throat, if any soreness at all, but as the cold gets underway will have noses that pour and pour and/or bad, lingering coughs.

****

Dr. William Bates, in his extensive study of vision problems, found that those afflicted with myopia could not, even when carefully covering their eyes to shut out all light, see an unrelieved black. Once the eyes could relax sufficiently to view, when closed and covered, unrelieved black, the myopic condition was cured.

Personality studies of myopes found that among other distinguishing traits they score low on proneness to guilt.

SPECULATIVE COMMENT:

Myopes aren't prone to guilt because they have lost the ability to see blackness, ie, the blackness of their own sins.

To phrase it another way: myopes have sharply curtailed their vision to avoid seeing the blackness of their sins.

A CURATIVE APPROACH:

To brighten myopic vision, therapy should encourage myopes to "see" the extent to which all of us share guilt.

PREDICTION:

With a relaxed acceptance of guilt—acceptance of the "blackness" of the human condition—myopes should regain normal vision.

****

Studies could easily determine whether there is any correlation between repressed maternal feelings/repressed lesbian feelings and breast cancer, as was hypothesized herein, also whether a correlation exists between sexual identity problems, a deep seated rage over sexual stereotypes and sexual expectation, and cancer of the lung.

One final topic: bodily weight.

Americans are currently suffering, in almost epidemic numbers, from what most doctors consider unhealthy: overweight. Diet pills and diet books sell by the millions, yet most of us are still bothered by an inability to keep our weight down to where we would like it. Why?

What is the etiology of this national malady?

We are an affluent nation, of course, with rich, fattening goodies readily available. But why, when we want to be slim and trim, do we stuff our faces with calories just because they are there? Is there some underlying reason that in our hearts, no matter what we think, we really want to be fat?

Consider this: If you had an extremely important paper that you had to place on a desk, a paper that you absolutely did not dare lose, how would you guard it? Keep it from flying away, becoming misplaced? Chances are that you'd carefully weigh it down with something.

Imagine this: You are in the kitchen of a house that has an underground basement. The door to the basement opens into the kitchen where you are. You know that a dangerous maniac is prowling around in the basement below and that if this maniac climbed the stairs and came through the door, he would do you harm. What to do? Well, why not move the refrigerator over against the door? The weight of that should keep the door closed and the maniac out.

SPECULATIVE COMMENT:

If we are repressing an enormous amount of anger/fear that we are afraid might climb the steps, push through the door and enter the conscious mind, one way to feel safer is to barricade the door with a great deal of weight.

If we are so frightened of our own negative feelings that we don't dare let them see the light of day, to ease our tension we can bury these feelings under layers of fat. That way they won't be able to get us. We might not like our weight, but it's still easier than worrying that the maniac prowling around in the basement below will burst through the door and wreck God knows what havoc. Anything is better than living with the terror of that.

DIETARY SUGGESTION:

Those who sincerely want to lose weight might risk peeking down into the basement just long enough to lose some of their fear of their own highly charged negative feelings. As they gain the courage to let those feelings sneak through into the conscious mind little by little, their need to reassure themselves with love—food—will decrease and their weight will drop off. Anger is a monster, but a monster we all have living within us. Better to tame those feelings, if possible, and allow them to live upstairs than to chain them down in the basement and live in terror of their escape.

SUMMATION:

Our knowledge of the universe comes to us through our five senses, senses which have proven to be limited and which frequently deceive us. We are equipped to see radiation only within a certain spectrum. Waves produce sounds above and below the vibratory speed which we are able to detect. Our fingers, touching numerous surfaces, tell us these surfaces are solid when in fact material objects are never "solid," but instead are composed of bits of mass whirling at dizzying speeds through vast, empty spaces.

An inner voice tells each of us that he or she is a distinct and separate entity, while our senses—limited and deceptive—tell us that the bodies we inhabit are solid and "real," that is, that they have an independent reality.

This book contests both of these experiential "facts."

Unique each of us is, but unique as a part of the whole, neither separate nor distinct, and our bodies are no more independently "real" than they are solid.

Instead:

Just as, on one level of being, the physical body casts or seems to cast a shadow, on another level of being *the mind casts a shadow of itself and that shadow—which has no independent reality—is the physical body.*

To change the configuration of the shadow-body we must change the shape of the mind of which the body is a shadow.

A creative force *does* exist, not as a clone of man, coalescing into a white bearded patriarch-God of the sky, but in man himself, if nowhere else.

Man comes into life endowed with godlike powers to create himself: his body, his mind, his heart, his life.

Man, a creative force, is in essence a spirit.

Hail to thee, brave spirit.

Hold to hope.

Flow with ease.

Create with love.

NOTES

Chapter 3:

1. L. A. White, "Erotica and Aggression: The Influence of Sexual Arousal, Positive Affect, and Negative Affect on Aggressive Arousal, Positive Affect, and Negative Affect on Aggressive Behavior," *Journal of Personality and Social Psychology,* 1979 (Apr), Vol. 37 (4), 591-601, as summarized in *Psychological Abstracts,* Vol. 64, No. 4, Oct., 1980, Pg. 871, Abstract No. 8021.
2. W. A. Rossi, *The Sex Life of the Foot and Shoe,* Saturday Review Press: E. P. Dutton & Co., Inc., New York: 1976, pg. 54.
3. Ibid., pg. 55.
4. Ibid.

Chapter 4:

1. C. Stern, Principles of Human Genetics, Third Edition, W. H. Freeman and Co., San Francisco: 1973, pg. 22.
2. P. Goldstein, *Genetics is Easy,* Fourth Edition, Viking Press, New York: 1967, pg. 11.
3. Ibid.
4. Ibid.
5. Ibid, pg. 42
6. Ibid., pg. 43
7. Ibid.
8. Ibid., pg. 18.
9. Ibid.
10. Ibid.
11. Ibid., pg. 21.
12. Ibid.
13. Ibid.
14. Ibid., pg. 22.
15. Ibid.
16. C. Stern, op. cit., pg. 41.
17. Ibid., pg. 90
18. P. Goldstein, op. cit., pg. 103.
19. Ibid., pg. 104.
20. Ibid.
21. Ibid., pg. 106.
22. Ibid., pg. 107.
23. Ibid.
24. Ibid., pg. 108.
25. Ibid., pg. 268.
26. Ibid.
27. Ibid., pg. 267.

Chapter 5:

1. P. Goldstein, op. cit., pg. 242.
2. Ibid., pg. 246.
3. Ibid., pg. 248.
4. Ibid., pg. 242.
5. G. Chedd, "Genetic Gibberish in the Code of Life," *Science 81,* Vol. 2, No. 9, November, 1981, pg. 54.

6. Ibid.
7. Ibid., pg. 51.

Chapter 6:

1. C. Eron, *The Virus That Ate Cannibals,Six Great Medical Detective Stories,* Macmillan Publishing Co., Inc., New York: 1981, p. 77.
2. Ibid., pg. 74.
3. Ibid.
4. Ibid.
5. Ibid., pg. 75.
6. Ibid.
7. Ibid., pg. 76.
8. Ibid.
9. Ibid., pg. 97.
10. Ibid., pg. 77.
11. Ibid., pg. 84.
12. Ibid., pg. 87.
13. Ibid., pg. 88.
14. Ibid., pg. 89.
15. Ibid., pg. 91.
16. Ibid., pg. 96.
17. Ibid., pg. 97.
18. Ibid.
19. G. Williams, *Virus Hunters,* Alfred A. Knopf, New York: 1959, pgs. 24-25.
20. Ibid., pg. 28.
21. R. Reid, *Microbes and Men,* Saturday Review Press: E. P. Dutton & Co., 1975, pg. 18.
22. Ibid., pg. 10.
23. Ibid.
24. Ibid., pgs. 10-11.
25. Ibid., pg. 11
26. Ibid.
27. *Encyclopaedia Britannica,* 14th edition (1945), Vol. 22, pg. 817c.
28. Eron, op. cit., pg. 14.
29. Ibid.
30. Ibid., pg. 15.
31. Ibid.
32. Ibid., pg. 17.
33. Ibid.
34. Ibid., pg. 18.
35. Ibid.
36. Ibid.
37. P. de Kruif, *Microbe Hunters,* Harcourt, Brace & Co., New York: 1926, pg. 315.
38. Ibid.
39. Ibid.
40. Ibid., pg. 317.
41. Ibid., pg. 317-318

42. Ibid.
43. Ibid.
44. Ibid., pg. 319.
45. F. Eberson, *Microbe's Challenge,* Ronald Press, 1963, pg. 311.
46. P. de Kruif, op. cit., pg. 325-326.
47. Ibid., pg. 326.
48. Ibid.
49. C. Eron, op. cit., pg. 28.
50. P. de Kruif, op. cit., pg. 333.
51. C. Eron, op. cit., pg. 28.
52. P. de Kruif, op. cit., pg. 323.
53. Ibid., pg. 332.
54. C. Eron, op. cit., pg. 31.
55. Ibid.
56. Ibid.
57. Ibid.

Chapter 7:

1. *Encyclopaedia Britannica,* 14th edition (1945), Vol. 20, pg. 816b.
2. Ibid.
3. Ibid, 816c.
4. *The Pocket Dictionary of American Slang,* complied by H. Wentworth and S. B. Flexner, Pocket Books, New York: 1968, pg. 74.
5. Ibid., pg. 175.
6. *Encyclopaedia Britannica,* 14th edition (1945), Vol. 20, pg. 816b.
7. C. Eron, op. cit., pg. 16.

Chapter 8:

1. "Medical Information on Myopia (Nearsightedness)," *American Association of Ophthalmology,* 1969.
2. Ibid.
3. J. Rodale and Staff, *Encyclopedia of Common Diseases,* Rodale Books, Inc., Emmaus, Penna., 1973, pg. 57.
4. Ibid.
5. "Medical Information on Myopia," op. cit.
6. Ibid.
7. Ibid.
8. J. I. Rodale and Staff, op. cit., pg. 57.
9. Ibid., pgs. 57-58.
10. W. H. Bates, *The Cure of Imperfect Sight by Treatment Without Glasses,* Central Fixation Publishing Co., New York City: 1921, pg. 89.
Reprinted as: *The Bates Method for Better Eyesight Without Glasses,* Holt, Rinehart and Winston, New York, 1940 (pg. 35).
11. F. A. Young, R. M. Singer and D. Foster, "Psychological Differentiation of Male Myopes and Nonmyopes," *American Journal of Optometry and Physiological Optics," 1975 - Oct., Vol. 52(10), pgs. 679-686,* as summarized in *Psychological Abstracts,* Vol. 55, No. 1, Jan. 1976, pg. 755, Abstract No. 7478.
12. R. M. Singer, "The relationship between visual refractive error and interests, values, personality, academic performance and reading," as reported in *Dissertation Abstracts International,* B-Sciences, Jul. Aug. 78, No. 8933, pg. 1019B.

13. Ibid.
14. Ibid.
15. "Medical Information on Myopia," op. cit.
16. C. A. Zeiger, "A Psychological Approach to the Improvement of Myopia," as reported in *Dissertation Abstracts International,* B-Sciences, June 77, Nos. 11-12, pg. 6359A.
17. W. H. Bates, op. cit., pg. 3.
18. A. Freese, MD, *Help for Your Arthritis and Rheumatism,* New American Library, New York: 1978, pg. 3.
19. Ibid., pg. 9.
20. Ibid., pg. 4.
21. Ibid.
22. Ibid., pg. 48.
23. Ibid.
24. Ibid., pg. 49.
25. Ibid., pg. 50.
26. Ibid., pg. 51.
27. Ibid., pg. 48.
28. Ibid., pgs. 54-55.
29. Ibid., pg. 55.
30. Ibid., pg. 50.
31. Ibid., pg. 57.
32. Ibid.
33. Ibid., pg. 59.
34. Ibid., pg. 142.
35. Ibid., pg. 141.
36. J. I. Rodale and Staff, op. cit., pg. 337.
37. Ibid., pg. 338.
38. H. Geist, PhD, *The Psychological Aspects of Rheumatoid Arthritis,* Charles C. Thomas Publishers, Springfield, Ill: 1966, pg. 17.
39. Ibid., pg. 18.
40. Ibid.
41. Ibid., pg. 33.
42. Ibid., pg. 63.
43. Ibid., pgs. 75-76.
44. Ibid., pgs. 69-70.
45. Ibid., pg. 77.
46. Ibid., pg. 80.
47. A Freese, op. cit., pg. 81.
48. H. Geist, op. cit., pg. 81.
49. J. W. Varni, "Self-Regulation Techniques in the Management of Chronic Arthritic Pain in Hemophilia," *Behavior Therapy,* 1981 (Mar), Vol. 12(2), 185-194, as summarized in *Psychological Abstracts,* Vol. 65, No. 4, April 1981, pg. 1423, Abstract No. 13289.
50. J. A. Baldwin, "Schizophrenia and Physical Diseases," *Psychological Medicine,* 1979 (Nov), Vol. 9(4), pgs. 611-618, as summarized in *Psychological Abstracts,* Vol. 65, No. 1, Jan. 1981, pg. 359, Abstract No. 3339 - J. D. Cooper.
51. A Freese, op. cit., pg. 36.
52. Ibid.

53. Ibid., pg. 35.
54. Ibid., pg. 37.
55. Ibid., pg. 41.
56. Ibid.
57. Ibid.
58. Ibid., pg. 42.
59. Ibid., pg. 39.
60. Ibid.
61. Ibid.
62. Ibid., pg. 37.
63. Ibid., pg. 74.
64. Ibid., pg. 73.
65. Ibid., pgs. 78-79.
66. Ibid., pg. 75.
67. Ibid., pg. 76.
68. Ibid., pg. 75.

Chapter 9:

1. *Encyclopaedia Britannica,* 15th edition (1979), Vol II, pg. 503.
2. *Encyclopedia Americana,* 1981, Vol. 5, pg. 530.
3. Ibid., pg. 528.
4. Ibid.
5. L. Israel, MD., *Conquering Cancer,* translated by Joan Pinkham, Random House, New York: 1978, pg. 17.
6. Ibid., pg. 19.
7. Ibid., pg. 20.
8. Ibid., pg. 21.
9. Ibid., pg. 23.
10. Ibid.
11. Ibid.
12. R. E. Netterberg, MD, and R. T. Taylor, *The Cancer Conspiracy,* Pinnacle Books, New York: 1981, pg. 18.
13. Ibid., pg. 17.
14. Ibid.
15. Ibid., pg. 3.
16. Ibid., pg. 38.
17. Ibid., pg. 42.
18. *Encyclopaedia Britannica,* 14th edition (1945), Vol. 4, pg. 737.
19. R. E. Netterberg and R. T. Taylor, op. cit., pg. 45.
20. Ibid., pg. 42.
21. Ibid., pg. 46.
22. Ibid.
23. Ibid., pg. 44.
24. Ibid., pg. 47.
25. Ibid., pg. 55.
26. Ibid., pgs. 54-55.
27. Ibid., pg. 54.
28. Ibid.
29. Ibid., pgs. 55, 58.
30. Ibid., pg. 64.
31. Ibid., pg. 66.

32. Ibid.
33. Ibid., pg. 64.
34. J. I. Rodale and Staff, op. cit., pg. 451.
35. Ibid., pgs. 519-520.
36. Ibid., pg. 520.
37. Ibid.
38. Ibid., pg. 521.
39. Ibid.
40. Ibid.
41. M. Scarf, "Images That Heal," *Psychology Today,* Sept. 1980, pg. 37.
42. Ibid., pgs. 37, 39.
43. Ibid., pg. 45.
44. Ibid., pg. 46.
45. Ibid.
46. L. Israel, op. cit., pg. 35.
47. Ibid., pg. 36.
48. M. Scarf, op. cit., pg. 43.
49. L. Israel, op. cit., pg. 49.
50. M. Scarf, op. cit., pg. 43.
51. P. Donahue and Co., *Donahue, My Own Story,* Simon and Schuster, New York: 1979, pg. 34.
52. L. Israel, op. cit., pg. 51.

Chapter 10:

1. Eron, op. cit., pg. 93.
2. Ibid., pgs. 93-94.
3. Ibid., pg. 94.

Chapter 11:

1. Fritjof Capra, *The Tao of Physics,* Bantam Books, New York 1975, pgs. 117-118.
2. Ibid., pg. 124.
3. *The Pocket Dictionary of American Slang,* op. cit., pg. 31.